PRAISE FOR *Buzz*

Winner of the 2019 Pacific Northwest Book Award

"Vividly zinging....[Hanson] zips and waggles through fascinating journeys to meet fellow bee obsessives, reminding us that...we have brought trouble upon ourselves: 40 percent of the bee species are in decline threatened with extinction."
    —*New York Times Book Review*

"Hanson is an insightful observer of evolution, at his most elegant when digging deep into the science....[His] senses are, indeed, sharp when observing the natural world."
    —*Wall Street Journal*

"*Buzz* shines the most brightly...when Hanson's own adoration of bees comes through: he wanders around the landscape observing them and musing about their natural history in ways that light up the page and make the book a rewarding choice for readers keen on science and nature."
    —**NPR**

"According to Thor Hanson's *Buzz*, the relationship between bees and the human lineage goes back three million years....Both our world and our brains, it seems, have been profoundly shaped by bees."
    —*New York Review of Books*

"Charming....Hanson is an upbeat and often humorous guide....If you have time to read one book on what is happening with modern bees, you couldn't do better than *Buzz*."
    —*Science*

"Engagingly written, well researched, widely informative.... Hanson is a conservation biologist with an infectious curiosity."
    —*American Bee Journal*

"Celebrates the wide diversity of bee anatomy and behavior.... The storyline here is a sadly familiar denouement to many modern natural histories, a tale of pathogens, habitat destruction, pesticides.... [But] if there are a few sour notes in its closing bars, the prevailing buzz in Hanson's book is sweet, sweet music."
    —*Natural History*

"[A] lively look at bees. From exploring the insect's evolutionary beginnings to profiling celebrity beekeepers, Hanson reminds us that we should appreciate these complex creatures, which are vital to our way of life."
    —*Seattle Magazine*

"Timely and convincing."
    —*Quarterly Review of Biology*

"Entertaining.... The author's point is plain: we need to adopt an approach to insects and to their central role in the maintenance of all life."
    —*New Statesman*

"This book is an accessible and engaging read; particularly enjoyable are Hanson's descriptions of his own forays into the world of bees."
    —*Edible East Bay*

"Delightful.... Bringing to mind Bill Bryson's complicated, but engaging ability to intertwine nature, science, art, history and culture, Hanson weaves a similar spell about the world's 20,000 species of bees."
    —*Winnipeg Free Press*

"If you think the sting of declining bee populations won't affect you, think again....Hanson take[s] himself and readers on a journey of exploration."
   —*Bend Bulletin*

"Enjoyable.... Hanson's enthusiasm is infectious."
   —*Current Biology*

"*Buzz* provokes thought and promotes action."
   —*Coast Weekend*

"[Hanson] is a charmingly enthusiastic bee fanatic and his book is a pleasure to read."
   —*Daily Mail*

"Hanson... is surprisingly optimistic that we can reform and protect our bees, citing recent research and improved agricultural practices. In *Buzz*, he states his case while entertainingly recounting human-and-bee history and his own experiences with many bee species."
   —*Booklist*

"Lively and entertaining, Hanson's work introduces the world of bees—all bees, not just honeybees.... Of interest to farmers, gardeners, ecologists, and anyone concerned about bees and their impact on our food supply."
   —*Library Journal*

"This beautifully written natural history book, brought to us by a graceful and talented author, packs surprise after surprise with every turn of the page. Who knew bees were just evolved wasps? Or that ancient Egyptians ferried bees up and down the Nile to pollinate their crops? Don't pass this one up."
   —**Wendy Williams, author of** *The Horse*

"Thor Hanson is a magician at making entomology and taxonomy exciting, highlighting the fascinating world of bees. *Buzz* hums with science and history, exposing how bees have shaped our world. A delightful, buzzworthy must-read!"

—Daniel Chamovitz, author of *What a Plant Knows*

"As he did for feathers and seeds, Thor Hanson has written a wonderfully engaging work of natural history that will delight readers with its elegant prose, surprising stories, and deep humanity. Bees, so important to life on earth, are fortunate to have someone as passionate and knowledgeable as Hanson tell the tale of their evolutionary past, turbulent present, and precarious future. After reading *Buzz*, you will look at bees with a profound mixture of awe and gratitude."

—Eric Jay Dolin, author of *Black Flags, Blue Waters*, and *Leviathan*

"This book hums with the unique mixture of science, adventure, intelligence, wonder, linguistic virtuosity, and great storytelling we have come to expect from Thor Hanson's work. But it offers something new and rare as well. Here we are drawn into a surprising and enchanted world that is hidden in plain sight. All who read *Buzz* will eat dinner, walk in the neighborhood, search the flowers, and yes— listen to the drone of bees—with changed minds and hearts, ones that are freshly attuned to our beautiful and essential interconnection with the six-legged beings who share and co-create our history, our mythology, our sustenance, our planet, and our future."

—Lyanda Lynn Haupt, author of *Crow Planet* and *Mozart's Starling*

"This book is a joy. In it, Thor Hanson reminds us that the story of bees is the story of the origin of societies, of sweetness and collapse, of flowers and their sex, and if the humans who study all of these things. It is a story of evolution and biodiversity, a story that bears on much of the food we eat but also so very much else. Buy it. Read it. Read it again. And when you do, you will look out at the buzzing world anew."

—Rob Dunn, author of *Never Home Alone* and *The Wild Life of Our Bodies*

"Thor Hanson is a gifted story teller and naturalist. In *Buzz*, he takes us along on a wondrous, action-packed journey to discover the secret lives of bees, flowers, and the unconventional men and women who study them. This book really is the buzz about bees, and it's destined to become a natural history classic."

—**Stephen Buchmann, author of** *The Reason for Flowers*

BUZZ

ALSO BY THOR HANSON:

The Triumph of Seeds

Feathers

The Impenetrable Forest

FOR CHILDREN:

Bartholomew Quill

THOR HANSON

BUZZ

The Nature and Necessity of Bees

BASIC BOOKS

New York

For Noah

Basic Books
Hachette Book Group
1290 Avenue of the Americas, New York, NY 10104
www.basicbooks.com

Printed in the United States of America

Originally published in hardcover and ebook by Basic Books in July 2018
First Trade Paperback Edition: September 2019

Published by Basic Books, an imprint of Perseus Books, LLC, a subsidiary of Hachette Book Group, Inc. The Basic Books name and logo is a trademark of the Hachette Book Group.

The publisher is not responsible for websites (or their content) that are not owned by the publisher.

Print book interior design by Trish Wilkinson

The Library of Congress has cataloged the hardcover edition as follows:

Names: Hanson, Thor, author.
Title: Buzz : the nature and necessity of bees / Thor Hanson.
Description: First edition. | New York, NY : Basic Books, Hachette Book Group, 2018. | Includes bibliographical references and index.
Identifiers: LCCN 2018012761 | ISBN 9780465052615 (hardcover) | ISBN 9780465098804 (ebook)
Subjects: LCSH: Bees—Popular works.
Classification: LCC QL565 .H36 2018 | DDC 595.79/9—dc23
LC record available at https://lccn.loc.gov/2018012761.

ISBNs: 978-0-465-05261-5 (hardcover); 978-1-5416-9953-3 (paperback); 978-0-465-09880-4 (ebook)

LSC-C

10 9 8 7 6 5 4 3 2 1

Contents

Contents

The Future of Bees _____

Author's Note

Although honeybees make many appearances in these pages, I want to state right up front that this book is not specifically about them. There will be no detailed descriptions of the waggle dance, swarming, or their many other unique and fascinating behaviors for the simple reason that those topics have been well covered elsewhere. Writers dating back as far as Virgil, and including at least two Nobel Prize winners, have produced hundreds of excellent volumes focused entirely on honeybees. This book, in contrast, celebrates bees in general, from leafcutters and bumbles to masons, miners, diggers, carpenters, wool-carders, and more. Honeybees feature as part of that panoply, but in this story, as in nature, they must share the stage.

Also, at the risk of vexing my entomologist friends, I've chosen to use certain words informally in this volume. Any insect might be referred to as a "bug," for example, rather than just those found in the order Hemiptera. Technical terms that couldn't be avoided are included in a glossary at the end, where readers will also find an illustrated guide to bee families, a bibliography of helpful references, and a collection of chapter notes. I heartily recommend the notes. They're full of intriguing tidbits that fell just outside the flow of the narrative—things like nectar pirates, date honey, and how the fuzzy-horned bumblebee got its name.

Acknowledgments

While writing books may sound like a solitary affair, it can only happen with the help and support of many skilled people. As always, I am indebted to my fantastic agent and guide through the literary labyrinth, Laura Blake Peterson, and once again I've had the great fortune to work with T. J. Kelleher and his crack team at Basic Books, including Carrie Napolitano, Nicole Caputo, Isabelle Bleeker, Sandra Beris, Kathy Streckfus, Isadora Johnson, Betsy DeJesu, Trish Wilkinson, and no doubt many others behind the scenes. I'm grateful to all the scientists, farmers, orchardists, and other experts who shared their stories and explained their work to me—any mistakes in the descriptions herein are mine and mine alone.

I would also like to thank, in no particular order, the following generous people and organizations who have boosted this project along in various ways, with apologies to anyone inadvertently forgotten: Michael Engel, Robbin Thorp, Brian Griffin, Gretchen LeBuhn, Jerry Rasmussen, Jerry Rozen, Rigoberto Vargas, Laurence Packer, Sam Droege, Steve Buchmann, David Roubik, Connor Ginley, Butch Norden, Beth Norden, John Thompson, Seán

Brady, Carla Dove, William Sutherland, Sophie Rouys, Patrick Kirby, Günter Gerlach, Gabriel Bernadello, Anne Bruce, Sue Tank, Graham Stone, Brian Brown, Alyssa Crittenden, Gaynor Hannan, George Ball, Mike Foxon, the Lyminge Historical Society, Martin Grimm, Robert Kajobe, Derek Keats, Jamie Strange, Diana Cox-Foster, Scott Hoffman Black, Ann Potter, the San Juan Preservation Trust, Dean Dougherty, Rob Roy McGregor, Larry Brewer, Uma Partap, Eric Lee-Mäder, Matthew Shepherd, Mace Vaughan, the San Juan Island Library, Heidi Lewis, the University of Idaho Library, Tim Wagoner, Mark Wagoner, Sharla Wagoner, Dave Goulson, Phil Green, Chris Looney, Jim Cane, Cameron Newell, Kitty Bolte, the Xerces Society, Bradley Baugher, Baugher Ranch Organics, Jonathan Koch, Steve Alboucq, and Chris Shields.

Finally, I am ever thankful for the unwavering support and patience of my wife, my son, my extended family, and a wonderful community of friends.

Preface

A Bee in the Hand

Full merrily the humble-bee doth sing,
Till he hath lost his honey and his sting.

—William Shakespeare
Troilus and Cressida (c. 1602)

The crossbow fired with a dull thwack and we watched its bolt disappear upward into the leaves and branches, trailing a length of monofilament fishing line that glinted in the scattered beams of sunlight. My field assistant looked up from the bow's sight and nodded in satisfaction, feeding out more line from a spinning reel duct-taped to the front grip. For him, this was all in a day's work, the standard procedure for helping biologists position ropes and research equipment high up in Costa Rica's rainforest canopy. For me, it marked a turning point. Within minutes, a colleague and

I had hoisted our insect trap into position, and for the first time in my career, I was officially studying bees. Or at least trying to.

The project didn't exactly go as planned. Days of shooting arrows at trees and hauling up various contraptions produced only a handful of specimens, mostly from one exciting moment when a dangling trap knocked into a nest and the whole hive attacked. The situation was infuriating—not only for all the wasted time and effort, but because I *knew* the bees were up there. I could see them clearly in the reams of genetic data I'd collected on the very trees where we were setting our traps. By comparing DNA from the adult trees to that of their seeds, I knew that pollen was moving around all over the place—not just among neighboring individuals, but between trees nearly a mile and a half (2.3 kilometers) apart. And because those trees belonged to the pea family, I knew that their clusters of purple flowers were designed for bee pollination, just like the vetches, clovers, sweet peas, and other common varieties back home. In the end, I had to admit defeat, but the experience sparked a fascination that wouldn't rest. I immediately sought out courses on the taxonomy and behavior of bees, and I've looked for ways to chase after them—in my work and in daily life—ever since. Sometimes, I even catch a few.

Like anyone else interested in bees, I've followed recent trends with an increasing sense of worry. Since beekeepers first reported signs of "Colony Collapse Disorder" in 2006, millions of domestic honeybee hives have simply winked out. Investigators point to a variety of causes, from pesticides to parasites, and have also uncovered steep declines in many wild species. With news reports, documentaries, and even a presidential task force ringing the alarm bells, public awareness of the situation has perhaps never been higher. But what do we really know of bees? Even experts often stumble over the details. Once, while listening to the radio in my car, I heard a noted historian of science describing how early colonists arriving at Jamestown and Plymouth brought honeybees with them from Europe. If they hadn't, he explained, there would have been nothing to pollinate their crops. I nearly drove off the road! What about the *four thousand species* of native bees already buzzing happily around North America? But that's not the worst of it. On the bookshelf in my office, I keep a hardbound copy of the volume *The Bees of the World*. It was written by well-regarded entomologists and published by a good nonfiction press, and the cover features a lovely close-up photograph . . . of a fly.

It's often said that bees provide every third bite of food in the human diet, but like so many of the natural wonders we rely on, they now fly mostly under the radar. In 1912, British entomologist Frederick William Lambert Sladen observed, "Everyone knows the burly, good-natured bumble-bee." That may have been true in the English countryside of Sladen's day, but a century later we find ourselves more familiar with the plight of bees than with the bees themselves. I once conducted a study in the patches of seaside prairie that lie just down the road from my house. I had a small grant to help answer one of biology's most basic questions: What's out there? Because, although I live within a day's journey of six research universities in two countries, we still don't have a good list of the local bees. The forty-five species I collected that season are just a start. Luckily for all of us, reconnecting with bees can be as easy as walking out our front doors on a summer day, wherever we happen to live. Filter out the commotion of modern life, and you can still hear them buzzing—those ubiquitous but overlooked visitors to every patch of open ground, from orchards, farms, and forests to city parks, vacant lots, highway medians, and backyard gardens. It's also lucky that what we do know about bees makes for an irresistible story. It begins with ancient specimens trapped in amber and soon moves on to honey-loving birds, the origin of flowers, mimicry, cuckoos, scent plumes, impossible aerodynamics, and, quite possibly, a major step in our own evolution.

Bees today certainly need our help, but just as importantly, they need our curiosity. Exploring the history and biology of these essential creatures can transform anyone into an enthusiast, and that is the purpose of this book. But I hope you will do more than read it. I hope it makes you want to go straight outside on the next sunny day, find a bee on a flower, and settle down to watch. If you do, you might just find yourself daring to reach out and catch that bee the same way my young son has done since the age of three—bare-handed. Try this and you, too, can feel the tickle of tiny feet and the whispery rustle of wings on your palm before you slowly part your fingers, hold the bee up, and set it free.

The Buzz About Bees

To lie and listen — till o'er-drowsed sense
Sinks, hardly conscious of the influence —
To the soft murmur of the vagrant Bee.

—William Wordsworth
"Vernal Ode" (1817)

Nobody trusts an exoskeleton. The mere sight of insects and other arthropods can trigger a measurable fear reaction in the human brain. Often, synapses associated with disgust also light up. Psychologists believe these feelings are innate, an evolutionary response to something that might bite, sting, or transmit disease. But there is also a deep sense of otherness about those brittle, segmented bodies: even from a safe distance, we know that such creatures would give a sickening *crunch* if stepped upon. Mammals like us belong to the vertebrates, animals who all share the chaste trait of tucking their structural parts out of sight *inside* the body in the form of bones. Technically, putting the hard bits on the outside may be the better evolutionary strategy—arthropod species outnumber

1

vertebrates by more than twenty to one. But the fact remains that people find exoskeletons creepy, particularly since they so often go along with faceted eyes, waving antennae, and multiple, scrabbling legs. Filmmakers understand this, which is why Ridley Scott based the terrifying monsters in *Alien* on insects and marine invertebrates rather than puppies, and why the scariest creature in *The Lord of the Rings* was not a pig-like orc or a cave troll, but Shelob, the giant spider. Even trained professionals sometimes fall prey to this squeamishness. In his book *The Infested Mind*, career entomologist Jeffrey Lockwood confessed to abandoning his research—and transferring to the philosophy department—after the grasshoppers he was studying suddenly overwhelmed him in a teeming swarm.

Too often, our interactions with arthropods end in a swatting motion, or even a call to the local exterminator. When we do make exceptions, they usually involve bugs that don't really look like bugs—butterflies that dazzle us with brilliant, colorful wings; woolly-bear caterpillars trudging along cheerfully under their furry tiger stripes; or ladybird beetles, beloved for what can only be described as unmitigated cuteness. People like crickets, too, but probably because their musical chirps can be enjoyed from a distance on a summer evening, without ever having to actually see one. In economic terms, the silk moth is appreciated for its valuable fibers, and we owe the entire world production of shellac to a small Asian scale bug, but our attitude toward insects is probably best expressed by global spending on pesticides, which currently tops $65 billion a year.

In the context of this general unease, the human connection to bees stands apart. With large, protruding eyes, two pairs of membranous wings, and prominent antennae, they do not hide their otherness. Young bees writhe like maggots, and when they mature, some species can swarm by the tens of thousands, each individual capable of delivering a painful, venomous sting. They look, in short, exactly like the insects we are meant to be afraid of. Yet, throughout

FIGURE I.1. The human fear of arthropods features heavily in our storytelling, from biblical locusts to Kafka's beetle to the horrors pictured on these pulp magazine covers from the 1920s. WIKIMEDIA COMMONS.

history, in cultures around the world, people have overcome or set aside that fear to bond with bees: watching them, tracking them, taming them, studying them, writing poems and stories about them, even worshiping them. No other group of insects has grown so close to us, none is more essential, and none is more revered.

The human fascination with bees took root deep in our pre-history, when early hominins sought out the sugary blast of honey at every opportunity. As ancient peoples migrated around the globe, they continued searching for that sweetness, robbing the honeybees as well as scores of lesser-known species. Stone Age artists captured the practice in cave paintings from Africa to Europe to Australia, depicting hunts that sometimes involved tall ladders, flaming brands, and dangerous ascents. To our ancestors, the value of honey justified effort and risk far beyond the inconvenience of a few pesky stings.

From raiding wild colonies, the transition to organized bee-keeping came as a logical next step nearly everywhere people settled down to farm. Potsherds laced with beeswax have been recovered from dozens of Neolithic agricultural sites across Europe, the Near East, and North Africa, some dating back more than 8,500 years. Exactly when and where the first beekeeper hived a swarm remains

FIGURE 1.2. Bees, hives, and people have appeared together in rock art for millennia, sometimes in literal depictions of honey hunting, but also in symbolic form, as in this ecstatic swarm and dance sequence from the San people, Eastern Cape Province, South Africa. IMAGE © AFRICAN ROCK ART DIGITAL ARCHIVE.

unclear, but Egyptians had certainly perfected the art by the third millennium BCE, tending their bees in long clay tubes, and eventually learning to ferry them up and down the Nile in concert with seasonal crops and wildflower blooms. People kept bees long before they tamed horses, camels, ducks, or turkeys, not to mention familiar crops like apples, oats, pears, peaches, peas, cucumbers, watermelon, celery, onions, or coffee beans. Domestication occurred independently in places as far-flung as India, Indonesia, and the Yucatan Peninsula, where Mayan beekeepers had the good sense to tend "royal ladies," a rainforest species with the agreeable trait of lacking a sting. By the time the Hittites ruled western Asia, beekeeping was enshrined in law, and anyone caught pilfering hives could expect a harsh fine of six silver shekels. The Greeks enacted honey taxes, required 300-foot-wide buffers between rival apiaries, and saw the trade become so lucrative that it inspired sophisticated counterfeiting. Herodotus described a convincingly syrupy substitute crafted from "wheat and the fruit of the tamarisk." Over the centuries, sticky liquids boiled down from dates, figs, grapes, and various tree saps would provide cheaper alternatives, but honey remained the world's ultimate measure of sweetness until the advent of refined sugars.

What began as an upshot of our primeval sweet tooth only grew stronger as people found other uses for the products of the hive. Mixed with water and fermented, honey soon provided the additional enticement of tasty and reliable intoxication. Scholars consider mead one of the oldest alcoholic beverages; it has been brewed and consumed in various iterations for at least 9,000 years, and perhaps far longer. Tipplers in ancient China quaffed a version laced with rice and hawthorn berries, while the Celts flavored theirs with hazelnuts and the Finns preferred the zest of lemons. In Ethiopia, people still favor a version doctored with the bitter leaves of buckthorn. But perhaps the most potent meads of all were those that *honey as medicine*

arose in the rainforests of Central and South America, where the Mayans and various tribal shamans developed hallucinogenic varieties spiked with narcotic roots and bark. In fact, healers of all kinds have long recognized the benefits of bees, recommending honey, mead, waxy salves, propolis (or "bee glue," a resinous substance collected from plant buds by some bees for use in hive construction), and even the venom from stings to treat all manner of ailments. When remedies from the ancient world were summarized in the twelfth-century Syriac volume *The Book of Medicines*, over 350 of its 1,000 prescriptions required bee products. The anonymous author went so far as to call honey water an essential daily tonic (when properly mixed with wine and one dram each of anise seed and crushed peppercorn).

The historian Hilda Ransome did not exaggerate when she wrote, of bees, "It is impossible to over-estimate their value to man in the past." As if sweetness, inebriation, and healing weren't enough, bees also provided nothing less than illumination. From prehistory through the dawn of the Industrial Age, most options for holding back the darkness involved no small amount of smoke and splutter—campfires, torches, or simple lamps and rushes that reeked of fish oil and animal fat. For all that time, only beeswax burned with a clean, steady, pleasant-smelling light. Temples, churches, and wealthy homes glowed with it night after night for millennia. Added to the many other uses for beeswax—from waterproofing to embalming to metallurgy—candle-making created an insatiable demand that often made wax the most valuable beekeeping product of all. When the Romans finalized their conquest of Corsica in the second century BCE, they spurned the island's famous honey in favor of a tribute measured out in wax alone—an impressive 200,000 pounds of it every year. Fittingly, the scribes and officials who oversaw that levy almost certainly made their notes on yet another bee-dependent innovation: the world's first conveniently erasable

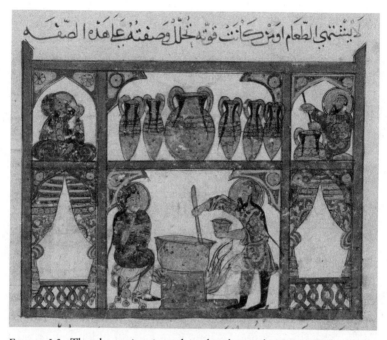

FIGURE I.3. The pharmacist pictured in this thirteenth-century Arabic text is mixing a typical cure-all for weakness and lack of appetite from a recipe that called for honey, beeswax, and human tears. Abdullah ibn al-Fadl, *Preparing Medicine from Honey* (1224). IMAGE © THE METROPOLITAN MUSEUM OF ART.

writing surface. Long before the invention of chalkboards, small tablets covered with wax could be inscribed with a stylus, easily stored or transported, and then heated, smoothed, and used again.

Bees have been with us from the start. As the source of so many commodities, some of them great luxuries, it's no wonder these insects found their way into folktales, mythology, and even religion. Bees in legend often played the role of messengers from the gods, their gifts to us seen as glimpses of the divine. Egyptians viewed them as the tears of the sun god, Ra, while an old French tale credits bees to Christ, formed from a scattering of droplets that fell from his hands as he bathed in the Jordan River. Deities and saints from

Dionysus to Valentine became patrons of bees and their keepers, while in India, bees made up the humming bowstring of Kama, the god of love. Swarms of bees often portended battles, droughts, floods, and other great events throughout the ancient world, symbolizing good luck in China, or bad luck in India and Rome. According to Cicero, a swarm of bees foretold the eloquence and wisdom of Plato by gathering and settling upon the lips of the philosopher when he was still a babe in arms. Bee-priestesses, known as *melissae*, Greek for "honeybees," served in the temples of Artemis, Aphrodite, and Demeter; they played a role at Delphi as well, where the famed Oracle was sometimes called "The Delphic Bee."

With its otherworldly sweetness, the syrupy diet of bees was also considered divine, appearing in legend nearly as often as the bees themselves. The mother of Zeus, for example, reportedly hid her infant son in a cave, where wild bees raised the young god to adulthood, passing sweet nectar and honey straight from their mouths to his. The Hindu deities Vishnu, Krishna, and Indra grew up on a similar diet and were known collectively as "the nectar-born ones," while in Scandinavia, the baby Odin preferred his honey mixed with milk from a sacred goat. Whether found in divine sippy cups or baked into heavenly cakes, honey dominated menus from Valhalla to Mount Olympus and beyond—traditions everywhere linked the sweetness gathered by bees to the foods of the gods. For the faithful, it also featured in the prospect of a just reward. Sources as varied as the Koran, the Bible, Celtic legends, and Coptic codices all described Paradise as a place flowing with rivers of honey.

In symbolism and in daily life, the value of bees to people lies rooted in their biology. The modern bee is a marvel of engineering, with wraparound ultraviolet vision; flexible, interlocking wings; and a pair of hypersensitive antennae capable of sniffing out everything from rose blossoms to bombs to cancer. Bees evolved alongside the flowering plants, and their most remarkable traits all developed in

FIGURE I.4. According to one Greek and Roman myth, this is where it all began, with Dionysus (Bacchus) capturing the first swarm of bees in a hollow tree. Piero di Cosimo, *The Discovery of Honey by Bacchus* (c. 1499). WIKIMEDIA COMMONS.

the context of that relationship. Flowers provide bees with the ingredients for honey and wax as well as the impetus for navigation, communication, cooperation, and, in some cases, buzzing itself. In return, bees perform what is their most fundamental and essential service. Yet, oddly, it's one that people didn't begin to understand—let alone appreciate—until the seventeenth century.

When German botanist Rudolf Jakob Camerarius first published his observations on pollination in 1694, most scientists found the whole notion of plant sex absurd, obscene, or both. Decades later, Philip Miller's description of bees visiting tulip flowers was still deemed too racy for his best-selling *The Gardeners Dictionary*. After numerous complaints, the publisher deleted it completely from the third, fourth, and fifth editions. But the idea of pollination could be tested by anyone with access to a farm, a garden, or even a flowerpot. Eventually, the dance between bees and flowers came to fascinate some of the greatest thinkers in biology, including such luminaries (and beekeepers) as Charles Darwin and Gregor Mendel. Today, pollination remains a vital field of study, because we know it is more than simply illuminating: it is irreplaceable. In the twenty-first century, sweetness comes to us from refined sugars, wax is a by-product of petroleum, and we get our light with the flick of a switch. But for the propagation of nearly every crop and wild plant not serviced by the wind, our reliance upon bees remains complete. When they falter, the repercussions make headline news.

Recently, the buzz *about* bees has often hummed louder than the bees themselves. Die-offs in the wild and in domestic hives threaten critical pollen and flower relationships that we've long taken for granted. But the story of bees is much more than a tale of plight or crisis. It leads us from the age of dinosaurs through an explosion of biodiversity that Darwin called an "abominable mystery." Bees helped shape the natural world where our own species evolved, and their story often comingles with our own. The subtitle of this book guides its content: it's an exploration of how the very nature of bees makes them so utterly necessary. To understand them, and ultimately to help them, we should appreciate not only where bees came from and how they work, but also why they've become one of the only insects to inspire more fondness than fear. The story of bees begins with biology, but it also tells us about ourselves. It explains why we've kept them close for so long, why advertisers turn to them to hawk everything from beer to breakfast cereal, and why our finest poets prefer their flowers "bee-studded," their lips "bee-stung," and their glades "bee-loud." People study bees to better understand everything from collective decision-making to addiction, architecture, and efficient public transportation. As social animals recently adapted to living in large groups, we have a lot to learn from a group of creatures who, in part at least, have been doing it successfully for millions of years.

In the past, people around the world heard the buzzing of bees as voices of the departed, a murmured conveyance from the spirit world. This belief traces back to the cultures of Egypt and Greece, among others, where tradition held that a person's soul appeared in bee form when it left the body, briefly visible (and audible) in its journey to the hereafter. While modern listeners perceive that living vibrato more prosaically, it remains a potent force, amplified by the unconscious urgency of a long and intimate bond. But the buzz about bees does not begin with pesticides, habitat loss, and the other challenges we've thrust upon them. It starts with their ascendance, an ancient lesson in hunger and innovation. Nobody knows the exact sequence of events that led to the beginning of bees, but everyone can agree on at least one thing: we know what it sounded like.

Becoming Bees

Evolution does not produce novelties from scratch.
It works on what already exists…

— François Jacob,
"Evolution and Tinkering" (1977)

CHAPTER ONE

A Vegetarian Wasp

You voluble,
Velvety,
Vehement fellows,
That play on your
Flying and
Musical cellos...

Come out of my
Foxglove; come
Out of my roses,
You bees with the
Plushy and
Plausible noses!

— Norman Rowland Gale,
Bees (1895)

I could not ignore the buzzing. My destination lay across the floor of a wide gravel pit, where I could see the white flutter of the rare butterfly I'd been hired to find. I should have been running toward it, net and notebook at the ready. But the ground at my feet hummed with an earthy tremolo that demanded immediate attention. This is the trouble with studying natural history—how to focus on any specific task when the world abounds with wonders. *Stay on target*, I told myself. That advice came to me from Luke Skywalker, who, during the chaotic final battle of *Star Wars*, struggled to keep his aim on the one tiny exhaust vent that would blow up the Death Star. Unfortunately for my clients, I lacked the concentration of a Jedi knight. The butterfly would have to wait.

Crouching down, I found myself surrounded by wasps, thousands of them. Their sleek black and gold bodies darted and swerved in every direction like sparks over a bonfire. But unlike sparks, the wasps eventually came to ground with purpose, landing beside the tiny nest holes that made up their colony, the largest I'd ever seen. I felt a surge of adrenaline, not from the danger of stings but from the thrill of discovery. For someone interested in bees, finding the right wasp nest can be like stepping back in time. If I was correct, the tiny burrows in the earth around my feet held a critical clue about how and why bees evolved. Putting net, notebook, and all thought of butterflies aside, I lay down with my face at ground level and began to watch.

A wasp immediately landed in the pebbly soil several inches away, moving back and forth with jerky motions almost too fast for the eye to follow. Homing in on a particular patch of sand, she suddenly stopped, thrust her front legs forward, and began to dig, flinging the spoils back through her hind legs just like a dog, or a tiny football player practicing the shotgun hike. Other wasps repeated this sequence all around me, their constant showers of flung sand making the ground appear to quiver. Some were tending old burrows, while others started afresh, but every individual worked apart. Unlike hornets, yellow jackets, and other more familiar wasps, these furious little diggers did not build elaborate paper nests or make pests of themselves at picnics. Nor did they live in large, organized communities led by a queen. They were solitary creatures, gathering en masse solely to exploit the benefits of a good patch of habitat. I recognized them as members of a diverse family still widely known by a name bestowed in 1802, the *sphecid* wasps.* This title comes straight from the word *sphix*, Greek for wasp, which means that to early entomologists, these insects embodied the wasp lifestyle so perfectly they deserved the formal description "waspy wasps." But the aspect of sphecids that had me face down in the sand dates back much further than Linnaean taxonomy. Sometime in the mid-Cretaceous, near the peak of the age of dinosaurs, a bold group of sphecids gave up one of their waspiest habits. Soon after, they evolved into bees.

In front of me, the individual I was watching suddenly stopped digging and flew off. Looking closely, I saw that she'd uncovered part of a burrow, her own or somebody else's I couldn't know. I waited for a few moments, but the wasp didn't return. So I reached out and began to brush away the sand myself, revealing a pencil-thin tunnel that angled slightly downward. Its walls began collapsing inward as I dug, so I inserted a long stem of dried grass to act as a guide. A few inches below the surface, grass and tunnel ended in a chamber that held just what I'd been hoping to find: the body of a fly. It was black and unremarkable, like something you might sweep off the windowsill on a summer day. But that one dead fly revealed something defining about the waspy wasps: they were hunters, constantly scouring the landscape for prey to feed their young. The species at

---

*Taxonomists recently divided the sphecid wasps into three families, putting the closest bee relations into a group called Crabronidae. Further revisions are expected, however, and the traditional, inclusive name used here remains common.

FIGURE 1.1. A colony of sphecid wasps in the genus *Bembix*, commonly known as sand wasps. Each female digs her own nest and brings prey to feed the growing brood within. Illustration by James H. Emerton, from George and Elizabeth Peckham, *Wasps: Solitary and Social* (1905).

hand, a type of sphecid called a sand wasp, concentrated on flies, but others took everything from aphids to butterflies to spiders, killing or paralyzing them with a sting and then depositing the bodies in a burrow to be devoured—dead or alive—by their growing larvae. The tactic is gruesome, but it's highly effective, a basic wasp strategy for over 150 million years. Changing it, however, proved to be even more successful.

Famous vegetarians from Leo Tolstoy to Paul McCartney have railed against slaughterhouses and promoted the various health and environmental benefits of a meat-free lifestyle. But campaigners continue to miss a great talking point when they omit the story of bees. For bees, vegetarianism did not simply alter their way of life, it created a new one. By making the dietary switch from animal parts to the sustenance provided by flowers, those first ancestral bees discovered an expanding and largely unexploited resource that was also extremely convenient. Where wasps typically needed to find one kind of food to feed themselves, and then track down something entirely different for their offspring, bees had the advantage of one-stop shopping. A good flower gave them sugary nectar for their own use, and protein-rich pollen that could be carried back home to nourish the young. And where flies, spiders, and other wily prey could be difficult or even dangerous to catch, flowers stayed put, and eventually began advertising their locations with alluring colors and scents. The exact details and timing of the transition from wasp to bee remain open to debate, but no one argues about how well it worked out. Bees now outnumber their sphecid relations by nearly three species to one.

After carefully refilling the burrow, I left the wasps behind and returned to my butterfly survey, spending the rest of the afternoon on a slope that glowed with blossoms—golden field mustard, red clover, and the purples of lupine and alfalfa. In the midst of such floral profusion, the idea of looking to flowers for sustenance seemed like a no-brainer. But in the world where bees evolved, it counted as nothing less than a chancy and pioneering adaptation. We think of the Cretaceous period for its dinosaurs, but reptilian profusion was hardly the only difference between that era and our own. The first bee to provision its young with pollen did so in a landscape where

FIGURE 1.2. Look past the fighting dinosaurs, and this scene reveals a typical impression of the mid-Cretaceous landscape—mossy, fern-ridden forests with no flowers or bees in sight. Illustration by Édouard Riou from *The World Before the Deluge* (1865).

there were no wildflower meadows as we know them, at a time when flowers themselves were still developing their petals, colors, and other characteristic traits. Fossils tell us that early flowers were tiny, inconspicuous things, bit players in a flora otherwise dominated by conifers, seed ferns, and cycads. Putting bee evolution in context requires a clear picture of that world, but most re-creations focus on the big lizards, not the vegetation. When I did look past the roaring beasts in dinosaur books, I couldn't find much of anything that looked like a flower, let alone a bee.

Struggling to visualize the *where* of bee evolution led me very quickly to questions about the *how* of it. If flowers were indeed small and rare in that world, then why would ancestral bees have sought them out? What sparked that vital vegetarian shift? What did the first bee look like? How long did it take to transition from wasp to bee? Whenever such questions arise about the evolution of insects, I've found it helps to call on the person who, quite literally, wrote the book.

"It's an amazing untold story for which we don't have much data," Michael Engel said, when I asked him about bee evolution. "To be crude," he went on, "the fossil record is piss-poor."

Michael spoke to me from his office in a warehouse owned by the University of Kansas. The school's insect collection (and its senior curator) moved there in 2006, when administrators decided that 5 million pinned specimens were taking up too much space in one of the grand old buildings on campus. He answered the phone with a curt "Engel," sounding like someone wearily accustomed to interruption. It's no wonder. In addition to his curatorial duties, he holds two university professorships, a research affiliation with the American Museum of Natural History, and editorial positions at nine different professional journals. His list of scientific publications includes more than 650 peer-reviewed articles as well as the credit that brought me to his doorstep, coauthorship of the definitive volume *Evolution of the Insects*. Within that wide-ranging topic, bees are his particular specialty. As soon as I reminded him that's why I was calling, his voice brightened and all other demands seemed forgotten. We talked for nearly two hours.

"To look for the earliest proto-bees you have to go back in time about 125 million years," Michael explained. Unfortunately, the oldest unequivocal bee doesn't show up in the fossil record until 55 million years later, leaving a huge hole right in the middle of the story. On the bright side, such a glaring lack of evidence may at least say something about *where* bees evolved. Because when fossils are particularly scarce, there's often a very good reason why.

"The sweet spot for the earliest bee is probably in the worst place for making fossils," Michael said. Several lines of evidence suggest that bees, as well as many early flowers, evolved in a dry, hot environment. Even today, the richest communities of bees lie not in the hyper-diverse wet tropics but in arid regions like the Mediterranean Basin and the American Southwest. Large parts of the Cretaceous landscape probably looked similar, but we know very little about those places, or about what lived there, because the making of fossils requires exactly what they lacked: water. To be fossilized, a creature or plant generally needs to be covered up quickly by sediments, preferably in a site starved of oxygen where it won't quickly succumb to rot. Such conditions occur primarily underwater, at the bottoms of swamps, lakes, rivers, and shallow seas. This means that our impression of the distant past, and our ability to study it, suffers from what paleontologists call "preservation bias." We're swayed by the flora and fauna from the wettest habitats, because, by and large, those are the things that turned into fossils. There are exceptions—fossils formed in dry places after flash floods or volcanic activity—but even these offer little help in sorting out the beginnings of bees.

"It's a conundrum," Michael told me. "You're stuck trying to find a fossil with the characteristics of a bee. But if you do, now it's a bee! You still don't know anything about the transition from wasps. You're damned either way."

The trouble lies in the very nature of what defines a bee: vegetarianism. Eating pollen is a behavior, not a physical trait, and behaviors don't make particularly good fossils. The tangible evidence for their new diet developed after the fact, with the evolution of distinctive hairs and other traits that helped them collect and carry pollen. (As long-haired, flower-loving vegetarians, bees have been jokingly referred to as "hippie wasps," which is actually not a bad way to remember their key evolutionary traits!) But the earliest bees must have looked just like their wasp relations, and may have continued that way for some time, perhaps carrying pollen in their stomachs and regurgitating it in the nest, the way some bees still do. This makes it highly unlikely that anyone will ever find the actual "first bee" (or recognize it if they happen to stumble across it).

"To really be sure, you would need a fossil nest," Michael mused. It would have to have pollen in it, preferably with the mama bee there, too, fossilized in the very act of provisioning. "And if anyone finds that," he added with a chuckle, "I will cash in my savings, buy a plane ticket, and fly to wherever they are in the world to see it!"

As we talked, it was obvious that Michael had a scientist's passion for data—and for making a clear distinction between ideas supported by evidence and those based on speculation. Bees are the vegetarian descendants of a sphecid wasp ancestor from the mid-Cretaceous. That much is known. Once we established that line, however, he obliged me by stepping across it and cheerfully entering the world of *maybe, what if,* and *perhaps.* As far as exploring the possibilities of early bee evolution, I could not have found a more qualified guide. "I'm one of the few people to waste serious time on it," he said wryly, though it's hard to call Michael's prolific output a waste. In 2009, the Linnaean Society honored him with their Bicentenary Medal, the most prestigious award in biology for scientists under forty. But if it hadn't been for a chance decision during his senior year of college, Michael Engel might never have looked twice at a bee in his life.

"I wasn't a bug kid," he recalled, though he'd always had an eye for detail. He liked to draw small things, and he'd drive his mother crazy insisting on expensive, extra-fine pens, so that he could get every feature exactly to scale. Later, he was firmly on the pre-med track at Kansas when a chemistry professor suggested he do something different for his honors thesis. "He said it would help my medical school applications stand out from the crowd," Michael explained. On a tip from his adviser, he wandered into the lab of legendary bee expert Charles Michener* and, in a sense, never left. The world of bee taxonomy fit perfectly with his love for getting the small things right, and he relished the challenge of solving difficult evolutionary mysteries. When I asked about his research approach, he described it this way: "If nobody else is studying something, then I want to study it." That contrary streak led him quickly to early bees, and insect evolution in general, when he heard a respected entomologist dismiss the entire insect fossil record as "not useful." After graduate studies at Cornell and a stint at the American Museum of Natural History, he returned to Kansas as Michener's hand-picked successor, inheriting a tradition of bee science dating back to the 1940s. Although he has published papers on everything from springtails and ants to termites, spiders, and a booklouse, bees and their evolution remain a primary focus. It's probably safe to say that Michael Engel has examined—and thought about—more bee fossils than pretty much anyone.

"My pet hypothesis," he told me, still in speculation mode, "is that wasps started fueling up on nectar, getting pollen on themselves incidentally, and then transferring it to the nest." It's also likely they started capturing prey on flowers—flies or other insects whose bodies may have been dusted with pollen, too, or who may have been eating it themselves. Either way, once pollen began arriving in the nest on a regular basis, the opportunity existed for wasp larvae to include it with the meat in their diets. And once that accidental

---

*The name and works of Charles Duncan Michener will crop up again and again over the course of this narrative. In a scientific career spanning eight decades, "Mich," as he was fondly known, established himself as the patriarch of bee studies. His books *The Bees of the World* and *The Social Behavior of the Bees* remain defining texts, and he trained scores of leading scientists, from Michael Engel and many other bee experts to the prominent population ecologist Paul Ehrlich.

delivery system became deliberate, it was, in Michael's words, a "run downhill" to the use of pollen exclusively.

"Suddenly, any female who spends more time on flowers avoids massive danger," he pointed out, noting the relative safety of pollen gathering compared to the risks of hunting. "Predation is a dangerous game. Prey will defend themselves, and if you get a tear in your wing, or you damage a mouthpart, you're in serious trouble." Natural selection would have immediately favored the pollen collectors, whose peaceable way of life helped them live longer and produce more offspring. "The next thing you know," he concluded, "you've got a bee."

Michael's scenario made a strong and intuitive case for the wasp-to-bee transition, but he was more circumspect about what happened next. Experts agree on the anatomical traits that define modern bees—even the most cryptic species share subtleties of wing venation and bear at least a few of the branched hairs so handy for transporting pollen. But the oldest known bee fossils already have these features, and the lack of anything earlier makes it impossible to know exactly when they evolved, or, in some cases, why. Even the origin of those telltale branched hairs is unclear, Michael pointed out. They might have evolved first to insulate flight muscles, or—if bees did indeed spend their formative years in deserts—to reduce water loss around their breathing holes. Until someone finds that perfect fossil nest that Michael dreams of, and a few more ancient bees to fill in the gaps, many of these questions will remain up for grabs. Fortunately, one needn't pin down the origin of every trait to get the gist of bee evolution. By the time they do start showing up as fossils, bees had clearly left their wasp ancestors behind to form a distinct, diverse, and highly successful group. And, as if to compensate for the earlier inconvenience, they appear in a form so beautiful that people have sometimes worn them as jewelry.

Michael's coauthor on the insect evolution book, David Grimaldi, once observed that his job required swinging two very different tools: a delicate net to capture living bugs, and a steel hammer to pry out the fossil ones. But even the hammer blows require finesse, particularly when the fossils are trapped in amber. Formed from the pitch of conifers and other resinous trees, amber deposits occur where ancient woodlands were flooded or otherwise covered rapidly by sediments. The fossilized resin ranges in color from its warm namesake to butterscotch, yellow, green, or even blue, making the excavation process something like prospecting for stained glass. But where glass is made for looking through, amber stands out for what can be seen within. Because any living thing trapped by its original sticky ooze gets preserved right along with the resin—not like the flattened outlines of a typical stone fossil, but in exquisite three-dimensional detail. Even microscopic features often show up clearly. In one famous case, a biting Cretaceous sand fly was so well preserved that its belly contained identifiable reptilian blood cells alongside known pathogens, providing evidence that dinosaurs, like people and other modern creatures, suffered the ravages of insect-borne disease.

For bees, amber provides the perfect medium, preserving all the fine anatomical details of a pollen-gathering lifestyle (and sometimes the pollen itself). Even in photographs, the fossils look startlingly lifelike and often quite beautiful, backlit and glowing in their translucent tombs. The oldest example, unearthed from a deposit in New Jersey that is also rich in flowering plants, dates back 65 million to 70 million years. The bee sits alone in a chunk of pale yellow amber, a female worker virtually indistinguishable from modern stingless species now common in the tropics. Just those basic facts, from a single specimen, show how far bees had already come. As honey-making hive builders with complex societies, stingless bees evolved only after more primitive, solitary species were well

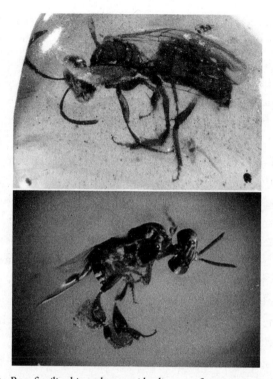

FIGURE 1.3. Bees fossilized in amber provide glimpses of extinct species in exquisite detail. This sweat bee (*Oligochlora semirugosa*, top) shows clearly visible wing veins, leg hairs, and antennae, while the stingless bee (*Proplebeia dominicana*, bottom), retains tidy balls of resin (collected for nest building) attached to its hind legs. Both specimens come from deposits in the Dominican Republic and are approximately 15 million to 25 million years old. TOP IMAGE COURTESY OF MICHAEL ENGEL VIA WIKIMEDIA COMMONS; BOTTOM IMAGE COURTESY OF OREGON STATE UNIVERSITY.

established. Simply finding enough pollen and nectar to support colonies of hundreds or thousands of workers required a flora already long accustomed to the presence of bees. Nearby plant fossils, from an even older forest, bear this out. They include ancient heaths with clumped pollen adapted to dispersal by fuzzy insects, as well as a relative of the genus of flowering plants called *Clusia* that appears to have produced resin in its flowers. This habit is thought to occur exclusively as a reward for highly specialized bees, who gather the resin to use in making their nests. Taken together, the New Jersey evidence proves that a lot happened between the time of the first bee and the time of the first bee fossil.

"It's a bit like showing up late to the party," Michael quipped, but even late arrivals can be fruitful. Before finding that fossil, experts could only speculate about the timing of bee evolution. Now it's clear that all the key steps, from physical features to social behavior, must have taken place early on. Bees may have started out as wasps, but they looked and behaved very much like they do today at a time when dinosaurs still roamed the earth. Unlike those ancient reptiles, bees appear to have more or less shrugged off the asteroid strike that brought the Cretaceous to a close. The most diverse fossil bee fauna known comes from a time shortly *after* that mass extinction, found in a source of amber so abundant that people harvest it with fishing nets.

Baltic amber formed in a vast European pine forest 44 million years ago and now occurs intermittently from northern Germany east as far as Russia. The most prolific deposits lie along the coast, including seabed veins that erode during winter storms and wash ashore to create what locals call the "season of amber." Gathered and traded since ancient times, this "gold of the north" was variously misidentified as petrified lynx pee, elephant semen, or the hardened teardrops of gods. Aristotle finally recognized its true nature, in part by studying the small creatures it sometimes contained. When Michael Engel turned his attention to Baltic amber, he found and described over three dozen bee species, including relatives of modern sweat bees, masons, leafcutters, and carpenter bees. Their appearance and variety fit neatly with the notion that bees evolved and diversified early, and they also coincided with a time when flowering plants were expanding quickly. But in addition to the scientific clarity they provided, reading Michael's papers left me with a nagging feeling about Baltic amber: I wanted to get my hands on some. Who could resist the idea of searching for ancient life inside a gemstone? Soon I found myself corresponding with a Latvian beachcomber who, for a small fee and the price of postage, agreed to send me a day's catch.

I live on a forested island in the Pacific Northwest, where it's easy to spot things getting themselves tangled up in tree pitch. A trail through the woods behind my office leads right past the oozing trunk of a Douglas fir, where I've watched ants, flies, spiders, beetles, a caterpillar, and three centipedes become irreversibly trapped and entombed. Locating insects—or anything at all—in a handful of beach-worn amber pebbles is another matter.

"Find any bees yet?" my wife asked with a smile. The packet from Latvia lay upended on the kitchen table, where I was busy with our young son, Noah, sanding and polishing the bloom off various chunks of amber and peering eagerly inside. Held up to the window, they glowed through with sunlight like brandy-colored jewels. We didn't find anything more than a few tiny splinters of wood and a knobbly bit that might have been a fleck of seed, but by the time Noah's interest began to wane, the kitchen was redolent with the smell of ancient pitch. Perhaps that was wonder enough, to breathe in the aroma of a long dead forest, still fragrant after 44 million years underground.

My amber collection now resides on a shelf beside my office window, where I also keep a selection of other fossils—Carboniferous leaves and seeds—and a replica of *Archaeopteryx*, the first bird. But it's the amber I go back to again and again, polishing and searching anew, particularly since noticing the scale bar alongside one of Michael's scientific illustrations. Where Noah and I had been expecting something blatant, like a bumblebee, the Baltic specimens were mostly tiny and inconspicuous, less than a quarter of an inch long

(< 6.5 millimeters). Many modern bees are just as small, making me wonder whether I would recognize one on a flower, let alone trapped in fossil resin. To really comprehend the diversity of bees, the sheer variety of their sizes and shapes and colors, I needed more than an insect net and a stack of books. I needed a guided tour. As it happens, just such a tour plays out every year at a remote field station in a landscape that—if Michael Engel's hunch is correct—looks pretty much like the place where the whole story of bees got started.

CHAPTER TWO

The Living Vibrato

Who knows not the names, knows not the subject.

—Carolus Linnaeus,
Critica Botanica (1737)

You never can tell with bees.

—A. A. Milne,
Winnie-the-Pooh (1926)

Two jet-black SUVs rumbled toward us down the dirt road, trailing clouds of dust that billowed and lingered in the dry desert air. They slowed to a stop, engines idling, and we could feel eyes watching us through the dark tinted windows.

"Oh, don't worry about them," Jerry Rozen said brightly, giving our invisible onlookers a wave. After decades of fieldwork in southern Arizona, he knew to expect a visit from the US Border Patrol. Mexico lay just half a mile to the south across a flat expanse that shimmered with August heat. But the figures moving around out there today had no interest in crossing international boundaries.

Instead, they dashed back and forth among the shrubs and cacti, swinging nets and calling out to one another whenever they made a good find. I was eager to join in, but first things first: a lesson in catching bees from one of the grand masters of the craft.

"Swing the net right over the tops of the flowers," Jerry instructed, demonstrating the proper technique with fluid back-and-forth strokes. Soon, his fine-meshed pouch hummed with a darting mass of angry insects. "Then you see what you've got," he added simply, and put the net over his head.

I don't know what was said inside the SUVs at that moment, but both vehicles suddenly revved their engines and sped off. Apparently, the Border Patrol had decided we were more of a threat to ourselves than to national security.

"The bees will always go to the light," Jerry continued, raising his voice a little from inside the net. Later he revised that statement to "almost always," and admitted to receiving the occasional sting right between the eyes. But today the insects cooperated, crawling up and away from his face as he lifted the tip of the net toward the sun. This gave him the chance to reach in with a glass vial and leisurely scoop out the bees that he wanted. Then he removed the net from his head and released the others with a backhand flip of the wrist. "Any questions?"

In the days ahead, everyone would have questions for Jerry Rozen. That was the whole point, the reason people had traveled from as far away as Japan, Israel, Sweden, Greece, and Egypt to attend The Bee Course. It provided a rare opportunity to brush up on bee biology—as well as to network and socialize—with some of North America's leading experts. Jerry's credentials included a stint at the Smithsonian followed by half a century (and counting) as the go-to bee curator at the American Museum of Natural History. Still nimble in his eighties, he carried himself with the grace and manners of an old-school naturalist, as smartly dressed for a day of

FIGURE 2.1. In this photo, a tiny golden bee in the genus *Perdita* perches on the huge black head of a carpenter bee in the genus *Xylocopa*. Both occur in Arizona, highlighting the incredible diversity of bees found in the deserts of the American Southwest. (Scale bar = 1mm.) PHOTO © STEPHEN BUCHMANN.

fieldwork as he was for evening gin and tonics on the porch of the research station. Jerry specialized in the nesting habits of hard-to-find solitary bees, while others on the faculty brought expertise in pollination ecology, genetics, and taxonomy. But the real focus of the course was something more fundamental: learning how to tell one bee from another. And few landscapes on earth offer a better place to do that than the deserts of the American Southwest.

When I first read the application, I thought there had been a misprint. Arizona in August? Who traveled to the desert during the hottest month of the year? But human comforts had little to do with scheduling The Bee Course. For bees, the heat meant perfect flying weather at a time when cacti and wildflowers were in full bloom, fed

by an annual flush of late summer rain. That combination made for
ideal habitat—abundant pollen and nectar in an otherwise parched
terrain that abounded with nest sites, from open ground and cut
banks for the diggers to hollow stems, rocky clefts, and rodent burrows for all the rest. With scant rainfall the rest of the year, those
nests rarely suffered the flooding, spoiled pollen, or fungal infections
that plagued bees in wetter climes. The resulting abundance meant
that any swing of our nets might conceivably yield an example from
over sixty different genera representing six of the world's seven recognized bee families. (For illustrated portraits of bee families, see
Appendix A.) To date, more than 1,300 species have been identified in Arizona, a diversity unmatched anywhere else on the continent. Soon, our days settled into an efficient pattern of classes and
collecting trips, followed by long hours in the lab preparing and
identifying specimens. With help from Jerry and others, I began to
recognize some of the major groups, mentally dividing the smooth,
black carpenter bees from the fuzzy bumbles, or the slender mining bees from the iridescent sweat bees and burly leafcutters. But on
that first day, when we all gathered for an evening lecture, the task
ahead looked downright impossible.

"Wrong! Not a bee!" Laurence Packer thundered gleefully,
and advanced the slide. To kick things off, he was challenging the
group's collective identification skills with a series of deceptively
wasp-like bees, bee-like wasps, and other confusing mimics culled
from his long experience studying tiny, cryptic species. He didn't
mean to discourage us, only to put our efforts into perspective. For
some bees, knowing the exact species required painstaking dissec-
tions, high-powered microscope work, and years of practice. But in
ten days, he assured us, we could learn to identify the broader taxonomic categories of family and genus. And since closely related bees
share behaviors as well as certain aspects of their appearance, those
skills would help us understand both the biology and the diversity

of bees at any site. But even with those caveats, Packer seemed extremely pleased whenever his pictures stumped us, particularly if he could fool his fellow instructors, too.

It was a fitting reaction. If Jerry Rozen was The Bee Course's elder statesman, then Laurence Packer was its provocateur. Standing over six and a half feet tall and dressed in a flowing cotton robe he'd picked up on a Middle Eastern expedition, he cut an imposing figure whether at the lectern or in the field. His opinions sometimes seemed outsized, too, but he matched them with a great patience—both for the bees and for those of us who were struggling to learn about them. When I joined him on a collecting trip the following day, we hurtled along the back roads in much the same way he talked—full speed ahead. But every time we paused to investigate a patch of flowers, he was genuinely eager to examine my catch.

"Well, you won't be needing those," he said at one stop, plucking three honeybees from my assortment and tossing them aside. Laurence spoke with the brisk cadence of his native England, though he'd spent his career in Canada, at Toronto's York University. Through lectures, books, and scores of scientific papers, he'd established a reputation for meticulous research coupled with a passionate defense of native bees. From Laurence I learned the term *melittologist*, Greek for "bee scientist." But he made a distinction between those who studied honeybees, a domesticated species, and those who studied bees in the wild. "It's not that I don't like *Apis mellifera*," he explains on his university website, using the honeybee's scientific name. But when people ask him questions about honeybees, he points out, it's like "asking an ornithologist a question about chickens."

Everyone I met at The Bee Course seemed to share Laurence's ambivalence. Whenever the topic of honeybees came up—and it always did—people spoke about them the way stage actors must talk about Hollywood stars, knowing that no amount of hard work on their part will ever bring the same level of fame. In spite of their diversity and importance, all the multitudes of wild bees stand in the shadow of their single, better-known cousin. It's sometimes a frustrating situation for those who study them. After all, outside of their native range in Africa, Europe, and Western Asia, honeybees often act like invaders, outcompeting native species and even introducing new diseases. But just as stage actors can still enjoy going to the movies, so can any melittologist appreciate honeybees. Many wild bee specialists are also active beekeepers, and I overheard lengthy debates about which flower nectars produced the tastiest honey. (Favorites included coffee blossoms, star thistle, and aromatic herbs like marjoram, thyme, and rosemary.) Honeybees also make good laboratory subjects, and we owe them much of what is known about bee anatomy, physiology, cognition, memory, flight dynamics, and advanced social behavior. So while they may be the chickens of the bee world, our industrious little domesticates have certainly earned their special status. Native bee enthusiasts like Laurence Packer simply want people to view honeybees as an *introduction* to bee diversity, not an alternative to it.

Personally, I felt grateful every time I caught *Apis mellifera* during The Bee Course. It pleased me to see them in my net for one reason: they have hairy eyeballs. Although experts don't agree on the function of the hairs (or even whether they have one), members of the genus *Apis* are among the very few bees so endowed—and honeybees are the only *Apis* bees in North America. I learned to spot those hairs at a glance and set their buzzing owners free without further thought. That meant fewer specimens to sort through in the lab, and, just as importantly, fewer bees that I had to kill. No matter how fond they may be of their subjects, melittologists must suffer the irony that studying bees often begins with the burnt almond reek of potassium cyanide, or the eye-watering fumes of ethyl acetate. Killing jars quickly produce piles of dead bees that must be fixed on pins and dried, their wings and legs carefully spread apart to show all the features necessary for an accurate identification.

I knew all of this going in. I understood the necessity and importance of scientific collections, and I knew that the vast majority of insect populations rebound quickly from the loss of a few individuals. But that didn't mean that I liked it. I've always felt a pang for the organisms my studies have led me to collect, even plants. It's a sentiment that would have limited my career prospects in earlier days. When Charles Darwin sailed on the *Beagle*, he shipped home everything from prickly pear cacti to a pickled hummingbird, more than 8,000 specimens in all. Alfred Russel Wallace was even more prolific in Malaysia, Indonesia, and New Guinea, where his total "specimens of natural history" topped 125,000. Modern biologists aim for a lighter touch, sampling with methods earnestly described as "noninvasive," or, even better, "sub-lethal." But for anything tricky to identify, taking a voucher back to the lab remained an essential step. I found that it helped to pretend I was fishing, and to start each collecting trip with a particular quarry in mind. One afternoon, midway through the course, I set out to catch what appeared to be a flying pearl.

The bee first came to my attention hovering over a coral pink cactus blossom, but I bungled the swing and snarled my net among the spines. It was a barrel cactus, with curved, dagger-sharp hooks that took quite a while to disentangle. This gave me the chance to glimpse the bee again, or another like it, when it stopped briefly at a flower nearby. Darting and quick, it had long, narrow eyes, a dark head, and a tapered abdomen banded with lustrous colors I couldn't quite define. For the next hour I stayed in the vicinity, but every netting attempt failed. I caught other things, but the bee of desire always kept just out of reach. Finally stopping for a break in the shade, I set down my net and took a long drink from my water bottle. Head tilted back, I spotted a familiar shape out of the corner of my eye. There was the bee, calmly taking its own rest right on the rim of my net! Thanking the fates for the bounty of a successful hunt, I scooped it straight into my kill jar and corked the lid.

That evening in the lab, my prize stood out from all the other specimens on the table. Up close I saw that its stripes weren't simply pearly, they were opalescent, flashing a rainbow of colors that shifted and swirled in the light. They looked gem-like because they produced their color in the same way that an opal did, not with pigments but through structure. When light strikes the surface of an opal, it diffracts and scatters through a glassy lattice of silica molecules, bending and separating into wavelengths our eyes perceive as colors. Those colors shift when we change our perspective to the waves, which is why any good jeweler will tilt an opal back and forth to show you the full glory of its glimmer. Remarkably, the bee's body was doing something very similar, scattering light not through silica but through a lattice of translucent chitin, the major component of its exoskeleton. The resulting hues graded from violet through blue to turquoise and on into green, yellow, and orange in a way that made it impossible to find the edge of any one color within the luminous whole. Even under magnification, the stripes were a hazy glow, their surfaces imprecise, as if the bee were made from light itself.

Happily, the evolution of opalescent chitin is as rare as hairy eyeballs, making my bee quite simple to identify. The trait occurs only in the alkali bees, a group named for their habit of nesting en masse in the mineralized soil of saltpans and dry lakebeds. The genus name, *Nomia*, comes from a beautiful mountain nymph known for seducing Greek shepherds. I could relate. While I've developed a great fondness for bees of all sorts, that *Nomia* was the first one I fell in love with. And while I've since encountered iridescent green and blue bees, bright red bees, and bees with plumy, snow-white fuzz,

I still consider *Nomia* the most beautiful. (It may be a good thing I've stayed true. Legend has it that Nomia the nymph once blinded a shepherd whose eyes and affections wandered.) During The Bee Course, I had no idea that I would one day stand in the buzzing midst of millions of alkali bees (an experience we'll reach in Chapter Five). Instead, I traveled home with my single treasured specimen and admired it so many times in the ensuing years that its head eventually fell off, prompting an emergency repair with Elmer's glue. It remains the bee I picture in my mind's eye as quintessential, the one I tag facts onto whenever I read about bee biology. So there can be no better example to turn to for the next part of this chapter, a guided tour of the remarkable anatomy of a bee.

For those of us accustomed to life with four limbs and internal skeletons, the bodies of bees seem utterly foreign. But there is an elegant logic to their architecture, each piece to its purpose, that may help explain why they're so wildly successful in nature. Like all insects, a bee consists of three basic parts: head, thorax, and abdomen. The head is for sensing and interacting with the world. It contains eyes, antennae, and mouthparts, everything a bee needs to see, smell, navigate, feed, and pick up things like pollen or nesting materials. Behind the head lies the thorax, the center of locomotion. Think of it as a big, armored muscle with attachment points for wings and legs, the essential tools of flight and crawling. From the thorax, the bee's body narrows briefly to a tiny waist before reaching the abdomen, the section so beautifully patterned on my alkali bee. Here lie the guts of the beast—all the organs and tubing needed for digestion, respiration, reproduction, and the circulation of blood. Scientists have been poking, prodding, and otherwise investigating the body parts of bees since at least the time of Aristotle, who observed that "the wing of the bee, if it is plucked off, will not grow again."

FIGURE 2.2. My quintessential specimen, the lovely alkali bee (*Nomia melanderi*). PHOTO © JIM CANE.

While whole books have been devoted to the subject, even the brief descriptions and stories below reveal volumes about how bees live, work, and perceive their world.

The head of my alkali bee resembles a small black lentil in size and shape, but one topped by two prominent antennae that rise up and arch backward from between the eyes. To continue the Arcadian theme, the antennae look like a pair of miniature shepherd's crooks pieced together from burls of smooth ebony. Among bee parts, antennae may be the most unfamiliar, because we really have no equivalent. Kids often call them "feelers," and it's not a bad name, since they're all about sensation. Imagine your nose perched on the end of a long, nimble stalk that also features taste buds, eardrums, and skin more sensitive than a fingertip. That would be something like a bee antenna, which sports seven distinct sensory structures, each one tuned to a particular environmental cue. Smell-ing involves microscopic pits and pores that constantly take in sam-ples from the surrounding air, allowing bees to sort through what one entomologist has called "a blizzard of odors." In a bee's world, chemicals signal everything from potential meals to potential mates, transforming any passing breeze into a tapestry of information. Like wine connoisseurs appreciating a complex bouquet, bees can easily tease apart the subtleties of a pheromone or pick out the fragrances of leaf, tree, soil, and water, all the while scanning for predators and the scent plumes of distant flowers. Antennae also process sounds and vibrations and play a key role in taste. They're covered with ultra-fine hairs and tiny pegs that respond to changes in tempera-ture, humidity, and airflow, while their tactile tips can distinguish the signature plushiness of various petals, from rose to aster to lark-spur. In the dark places where bees nest, antennae become the pri-mary means of navigation and communication, helping them find their way, find each other, and share scent-coded information about the work of the hive.

Had Aristotle plucked the antennae from his bee rather than the wings, he would have found the poor creature equally incapacitated. He would also have found himself in good scientific company. Ex-periments that trim, remove, or otherwise tamper with bee antennae are common, and they continue to discover new sensory abilities. Research now suggests that antennae influence body position in flight, respond to the earth's magnetic field, and pick up the faint electrostatic charges given off by flowers. The tiny distance between antennae—less than one-sixteenth of an inch (two millimeters) on my alkali specimen—is apparently far enough to sniff out minute differences in concentration between right and left, small sensory gradients that indicate the direction of an odor. Add just a few more scent molecules to the air on one side or the other, and the bee will turn to follow, an ability that allows it to track the drifting fragrance of a flower to its source from well over half a mile (one kilometer) away. Captive bees deprived of their antennae often appear disoriented, and have a hard time performing such basic tasks as landing on angled surfaces (i.e., flowers). While we can't know exactly what bees experience, we know they feel much of it through their antennae. Naturalist C. J. Porter marked this finding with apparent remorse in 1883, after clipping the antennae from a bumblebee. He said its apparent shock and stumbling confusion reminded him of an ox that had been "struck a hard blow on the horn," concluding, "I think it . . . fainted from the pain."

My alkali bee was long deceased and quite beyond pain when its head fell off, so I took the opportunity to peer in from the back side, hoping to glimpse the world through a bee's eyes. Unfortunately, dried tissues and chitinous struts filled up the whole interior, blocking all light and keeping the view from those great, elliptical orbs a mystery. It's often said that a bee has five eyes, but that is somewhat misleading. The extra three, called *ocelli*, poke up like glassy marbles from the top of the head, but amount to little more than light-sensitive nubs. Incapable of forming images, they appear to play a more limited role: tracking patterns of light intensity and polarization to help bees navigate, particularly at dusk. In terms of vision, the real action takes place inside the two enormous compound eyes that frame and dominate a bee's face. Each one contains over six thousand facets that constantly beam their individual views of the world to the brain, which knits all those pictures together into a single, wide-angle composite. Because the eye is rigid, however, its focal distance is fixed and quite short, making anything seen from afar look like a highly pixilated blur. Flowers, nest holes, fellow bees, and other objects of interest only resolve into sharp focus up close, within a few inches. This myopic view may seem limiting, but bees compensate with an extraordinary ability to sense motion. Every eye facet is individually hard-wired from lens to brain, which means that anything moving through a bee's field of view sets off not just one optic nerve, but a whole cascade of reactions, like a fingernail dragged across harp strings. Even the smallest motions stimulate scores or hundreds of facets, all of which glimpse that moving object from a slightly different angle. The result is a hyperawareness that also allows bees to unconsciously calculate speed, distance, and trajectory, which helps explain the many times my net has come up empty. (It also explains the much larger eyes of male bees, whose primary goal in life is to spot motion—the streak of an eligible female passing by on her nuptial flight.)

To the human eye—mine or anyone else's—the opalescent bands on an alkali bee shimmer with rainbows. Bees see the rainbows, too; they're just different rainbows. The visible spectrum for most bees starts somewhere in the yellower shades of orange, peaks in the bright blues, and continues on down into the short wavelengths known as ultraviolet. Although this removes red and maroon from their color vocabulary, it opens up a world of other possibilities. Ultraviolet light is known to people primarily as the source of sunburn, something to be blocked with long-sleeved shirts, thick lotions, and visored hats. We don't know what it looks like because, well, we can't see it. But cameras with special filters can tell us where it is, and they reveal a hidden language of attraction writ large on the petals of flowers. What we see as the uniform yellow of a dandelion, for example, appears differently to a bee—rich and luminous at the center, where yellow pigments combine with ultraviolet to produce a shade referred to as "bee purple." This combination and many others occur on the blossoms of more than a quarter of all flowering plants studied to date, and at a much higher rate for those visited by bees. Like the petals of a dandelion, ultraviolet colors on other flowers often create bull's-eye patterns or radial stripes called "nectar guides" that point like gleaming arrows toward their sources of sweetness and pollen. Those patterns are far

FIGURE 2.3. Picturing the ultraviolet colors seen by bees transforms our notion of many familiar flowers. Here, photographic filters reveal a rich wash of "bee purple" enhancing the bull's-eye pattern of a black-eyed Susan. The blossom is shown as it appears to the human eye (left) and as seen by a bee (right). PHOTO © KLAUS SCHMITT.

from arbitrary. A bee's-eye view of the world is driven by their near-constant searching for the flowers that sustain them. What happens when they find them, however, depends on other body parts, starting with the mouth.

The mandibles and tongues of bees look industrial, like things that should move with cogs and cables instead of muscle. They vary considerably in size and shape, depending on the need. A leafcutter bee, for example, has mandibles with fine, sharp teeth for snipping through greenery, while carpenter bees boast massive grinders for chewing wood. Honeybee mandibles look like spatulas, with broad, flat tips handy for spreading and shaping wax. Since my alkali bee was a ground nester, her mandibles doubled as shovels—mostly smooth and rounded, but with a single blunt tooth near the tip for prying and chipping hard earth. She held them crossed neatly below her chin like a pair of familiar tools, their edges well-polished by use. Below them, her tongue cantilevered out like a thin, coppery pipe, enameled black at the base and half again as long as her head. Bee tongues look solid but really consist of a central grooved and tufted shaft protected by overlapping sheaths. When the bee is feeding, muscles at the base flex a hollow bulb that works like a pump, quickly transferring nectar from flower to stomach. The whole apparatus is jointed, made to fold up inside the mouth cavity like the pleats of an accordion or the arm of an articulated crane. (Pinned specimens like mine feature the tongue deliberately extended for display.) Since the length of the tongue determines how far a bee can reach inside a blossom, some specialists have developed true colossi. In addition to his tricky wasp photos, Laurence Packer shared pictures of a species, yet unnamed, that he'd just discovered in Chile's Atacama Desert. Its tongue and elongated head stretched out like the trunk of an elephant, an absurd protuberance longer than the rest of its body, yet just right for reaching nectaries hidden in the depths of the borage flowers it fed upon.

Behind a bee's head lies the thorax, a collection of impossibilities. In the 1930s, French entomologist Antoine Magnan famously (and facetiously) implied that insect flight defies the laws of aerodynamics. Similar claims have been attributed to a German physicist and a Swiss engineer of the same era, and over time that notion has become inextricably linked to one particular insect, the bumblebee,

FIGURE 2.4. The bizarre, elongated head and tongue of this Chilean desert bee in the genus *Geodiscelis* evolved to help it reach the nectar located in deep flowers. PHOTO COURTESY OF USGS BEE INVENTORY AND MONITORING LAB.

whose fuzzy bulk looks too large for its wings. As a cultural meme, the "impossibility" of bumblebee flight is now a common allegory for attaining the unattainable, popping up everywhere from sermons to self-help books to political speeches. The eponymous founder of Mary Kay Cosmetics went so far as to adopt bumblebees as a corporate mascot, handing out diamond-studded bee pins to inspire a sales force of "women who don't know they can fly." Though it's true that a bee can't soar in the manner of a fixed-wing aircraft, it's also rather obvious that bee wings aren't fixed: they flap. Magnan and other early students of insect flight knew full well that the aerodynamics were different, but just how a bee's wings create lift remained mysterious until quite recently.

At rest on its pin, my alkali specimen holds its wings aloft as if frozen in midflight. Up close, they look like stained-glass windows

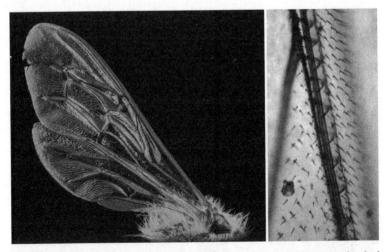

FIGURE 2.5. The paired wings on either side of a bee can be held apart or hooked together to function as one. The image on the left shows the small hind-wing and large forewing from the left side of a honeybee, joined by a row of hooks on the hind-wing that tuck into a fold on the trailing edge of the forewing. The connection is pictured in detail on the right. LEFT PHOTO COURTESY OF USGS BEE INVENTORY AND MONITORING LAB; RIGHT PHOTO © ANNE BRUCE.

awaiting the addition of color, their cellophane thinness strength-
ened by a lattice of dark, structural veins. Each side of the bee bears
two wings, though they often appear as one, held together by an
ingenious system of tiny hooks and folds. They look nothing like the
rigid, curve-topped wings found on airplanes, nor should they. Where
a fixed wing creates its lift through shape, angle, and airspeed, bee
wings fly as if by agility alone, flapping at a rate that often exceeds
two hundred strokes per second, and adjusting those motions to take
advantage of wind, air pressure, and the fickle vortices of their own
passage. The sheer speed of bee wings baffled early researchers—
such rapid-fire contractions seemed like another impossibility, faster
than the bee's brain could send signals to its nerves. But bees and
a number of other insects overcome this hurdle through elasticity
and the natural tension between opposing muscles in the thorax.
For every nerve impulse, those muscles vibrate like a plucked guitar
string, flapping the wings five, ten, or even twenty times before the
next impulse arrives. Just how those rapid strokes create lift only
came to light with the invention of high-speed video cameras able
to snap thousands of images in a second. Frame-by-frame analysis
showed the wings moving not up and down, as expected, but forward and backward, like a pair of sculling oars. Adding smoke to the
experiments illuminated airflow, revealing how rapid rotations and
adjustments in wing angle produce a steady downward pressure, like
helicopter blades, and also create whorls of low pressure that spiral
off the wings' upper surfaces, further increasing their lift. The resulting aerodynamic picture transforms our impression of bee flight from
anomaly to masterwork, a model for everything from drones to wind
turbines. Even the ungainly bumblebee has been redeemed, now
noted for its remarkable ability to remain airborne in thin mountain
air. A bumblebee species native to the Himalayas is thought to be
the world's highest-flying insect: it is still able to hover at elevations
beyond the peak of Mount Everest.

The earthly half of a bee's locomotion system juts and dangles from below the thorax in the form of six nimble legs. Less mysterious, perhaps, than wings, they remain no less remarkable. On my alkali bee, the legs are tiny and paper-clip thin, but under a microscope they leap into view like articulated steampunk machines. Unlike steampunk, however, where the elaborations are stylistic, every fringe, joint, and spike on a bee leg serves a purpose. Bending the forelegs, for example, closes a tiny spur over an opposing notch, forming a perfect circle just the right diameter for grooming antennae. Watch a bee before it leaves a flower, and you'll often see it reach up and pull its antennae through these gaps repeatedly, neatly removing any pollen or dirt that might impair its senses on the flight home. At the end of each leg, two curved, spine-like claws serve as feet, surrounding a fleshy-looking pad that works like a suction cup. This combination provides bees with traction as well as a gecko-like ability to cling to smooth surfaces. (The claws make it hard to shake a bee from your sweater; the pads make it hard to blow one off the rim of your glass.) My specimen had dried with one hind leg flung high up in the air like a dancer in a chorus line. To entomologists, that glitch would betray my relative inexperience as a bug-pinner, but it did show off a feature of the back legs that makes them particularly vital to the bee lifestyle. Even after years in storage, that leg glows with clumps of golden pollen, perhaps from the very cactus blossom where I first spotted it. The pollen stays put because it lies trapped in a dense fringe of finely branched hairs called a *scopa*. (Imagine trying to brush powdered sugar from a shag carpet, and you'll get the idea.) The other legs bear their own combs and brushes, used to gather pollen or pluck it from body hairs, before shifting it back to the scopa for storage and transport. Bumblebees, honeybees, and their close relations take the concept a step further, wetting the pollen with nectar so that it forms a sticky ball that can be tucked in a basket-like hollow shaped into the structure of the

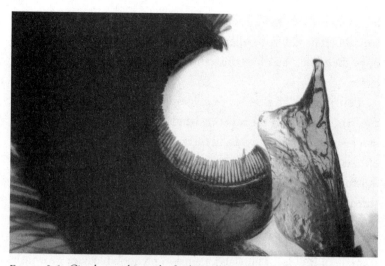

FIGURE 2.6. Circular notches in the forelegs of bees are perfectly sized for grooming antennae, as shown in detail here on the leg of a honeybee. PHOTO © ANNE BRUCE.

FIGURE 2.7. The hind legs of female bees often feature a dense fringe of branched hair for carrying pollen, like the shaggy fleece on this long-horned bee in the genus *Melissodes*. PHOTO COURTESY OF USGS BEE INVENTORY AND MONITORING LAB.

leg itself. If they visit various kinds of flowers on the same collecting trip, different-colored pollen types are often clearly visible, striped onto their back legs like the garish bloomers of an old-time circus clown.

Pollen aside, the center of coloration for most bees lies behind the legs, flashing on the tapering bands of the abdomen. Those hues can be built into the cuticle, as on an alkali bee, or displayed in tufts of hair tinted orange, yellow, black, white, or, in some tropical and Australian bees, bright blue. The colors often signal a warning—the threat of a sting—but they can also play a role in species and mate recognition, with males and females sometimes displaying different patterns. Bold stripes are common, but flamboyance isn't always the rule of the day. Many abdomens simply look black or brownish, and some probably glow with ultraviolet hues we can't perceive or classify. Color aside, the real work of abdomens occurs within, supporting the various organs and plumbing that keep a bee running. Most of these follow the standard insect model: a simple heart circulating blood to the brain and muscles, and a system of sacks and tubes drawing in and expelling air through tiny holes in the cuticle. Much of this activity happens passively, but when a bee exerts itself, it can speed things up by visibly pumping its abdomen, the insect equivalent of panting. The digestive tract of a bee stands out for the charmingly named "honey stomach," or "honey crop," a pouch that expands dramatically when needed, shoving other organs aside to make room for loads of nectar. Add in reproductive organs and a few glands for secreting pheromones and nest-building substances, and the abdomen is basically complete. There is one more feature at the back end of a bee, however, with the potential to make a lasting impression: the stinger.

If you ever find yourself seriously studying bees, or writing a book about them, the most common question people will ask you is how many times you've been stung. You can then surprise them with the

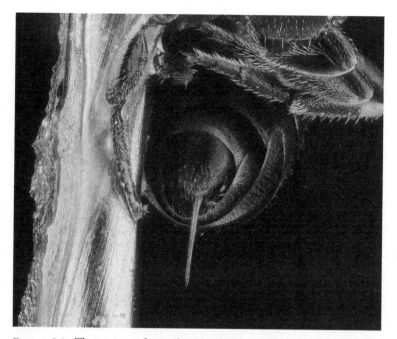

FIGURE 2.8. The stingers of most bees are barbless and needle-sharp, like this one on a small, masked bee in the genus *Hylaeus*, magnified here and shown next to the shaft of a pin for scale. PHOTO COURTESY OF USGS BEE INVENTORY AND MONITORING LAB.

news that most bees rarely sting at all, and some aren't even capable of it. Chief among these are the males, who lack the equipment entirely. Stingers evolved in the wasp ancestors of bees as an extension of the female reproductive system, from a pointy tube originally used to lay eggs. Only females have them, and only females can sting. For ancient wasps, this handy tool served dual purposes, first immobilizing a prey item and then laying eggs directly in or upon it, where their carnivorous larvae would hatch in a perfect place to feed. Many wasps still do exactly the same thing, but some groups, and all bees, eventually separated the two functions, diverting egg-laying activities to a small hole at the abdomen's tip, and devoting the tube-like stinger entirely to defense and attack. This allowed specialization to fit the lifestyles of particular bees, from entirely stingless species to those with a diabolical, self-pumping needle designed for group defense.

My alkali bee specimen died with her stinger extended in what must have been a final, defensive act. It looked like a tiny splinter jutting from her abdomen, but under magnification I could see that it consisted of several close-fitting parts: a grooved central shaft for delivering venom, flanked by two sharp lancets for piercing the flesh and holding. As in the vast majority of species, the lancets were smooth-edged, like amber stilettos, with just a few shallow jags near the tip for traction. This meant she could have easily withdrawn the whole thing to jab me more than once, which might have been a good idea on her part, since the sting probably wouldn't have hurt much. Though he didn't include the genus *Nomia* in his famous ranking of insect stings, entomologist Justin Schmidt likened the pain from a related bee to a tiny spark singeing a single arm hair. Without a large nest to defend, most bees only need enough potency to fend off the occasional rival, or the attack of a hungry spider. The real pain in the bee-sting world comes from larger, highly social species, whose nests contain multitudes of tasty larvae—and in some cases honey—making them attractive targets for everything from bears to birds to primates. In these species, worker bees employ group defense tactics to protect their nests from all comers. It's not just the quantity of venom that matters, but what goes into it—varying the proteins, peptides, and other compounds in the mix helps make a sting more toxic to its likely recipient. Mammals like us feel burning pain from a cell-destroying heart toxin called melittin, for example, while other insects (including other bees) are more affected by histamines.

Honeybees deserve special mention for equipping their lancets with barbs, wicked hooked teeth that fix firmly in flesh, keeping an inserted stinger attached to its victim. If a honeybee flies away or gets brushed off during an altercation, its stinger remains, tearing free from the abdomen with the venom sac and musculature still attached and pumping. The associated nerve center is also part of this package, enabling the stinger to "live" apart from the bee for over a minute, more than long enough to deliver its full dose of venom. For the honeybee, stinging results in a fatal abdominal wound, but with thousands of workers in any given hive, the benefits of a fearsome defense outweigh the loss of a few individuals. Schmidt considers the honeybee sting exemplary, a memorable baseline of pain useful for comparison with other insects. The most memorable description of that pain, however, comes from the Belgian Nobel Laureate and amateur entomologist Maurice Maeterlinck: "a kind of destroying dryness, a flame of the desert rushing over the wounded limb, as though these daughters of the sun had distilled a dazzling poison from their father's angry rays." Linking bees to the sun is fitting on many levels, and Maeterlinck's analogy concludes our tour of bee bodies more or less where it began, in the desert.

I left Arizona with my alkali bee and more than a hundred other pinned and labeled specimens in a cardboard box, a reference collection I still consult for help making identifications. The staff of The Bee Course pride themselves on teaching practical, hands-on scientific skills, but they can't help passing on something more, a contagious fondness for their subject. Knowing bees as an endearment enriches their study, changing the questions an observer might think to ask. Now that I can put a name to a bee, I can't help wondering what its life must be like as it flies through a world of different colors and constant motion, where vision interacts with memory, scent, vibrations, electrical charges, and magnetism to form a landscape of vivid sensation. I see a bee on a flower and now imagine how it got there, following a scent plume that swelled from faint whiffs to an intoxicating stream until the pixilated blossom finally came into focus, petals pulsing with bee purple, nectar guides, and a tingling, electric pull, leading the bee inevitably inward to its sweet reward. The bodies of bees are sophisticated machines for finding and transporting pollen and nectar, but the more I thought about their lives, the more I realized that something was missing.

I had caught my alkali bee in a patch of blooming cacti, and nearly every other bee in my collection was netted on or around a flower. For a bee hunter, what better place to look? But while visiting flowers is certainly central to the bee way of life, it's only part of what they do. Once they've filled their honey stomachs with nectar, and carpeted both scopa with pollen, where do they go? I knew that honeybees lived in hives of thousands, but I also knew that they were the exception. Most bees in my specimen box had led very different lives, building nests and raising their offspring completely alone, in ways I knew nothing about. If The Bee Course had lasted longer, I could have put my questions to Jerry Rozen, Laurence Packer, or one of the other instructors. But sometimes, if you want to hear a story, it's best to ask a storyteller. And I happened to know someone who once made his living by deciphering, packaging, and selling the story of solitary bees.

CHAPTER THREE

Alone Together

Solitude is certainly a beautiful thing, but there is pleasure in having someone who can answer, to whom one may say from time to time, that solitude is a beautiful thing.

—Jean-Louis Guez de Balzac,
On Retirement (1657)

At first, he didn't even realize they were bees. Brian Griffin was busy putting up a new garden gate when he noticed a few small black insects flying around the freshly dug post holes. He wondered briefly what they were up to, and then dismissed the matter from his mind. Recently retired after thirty-five years in the insurance business, Brian felt eager to get on with long-deferred projects and hobbies: woodworking, watercolors, local history, and gardening. Entomology wasn't even on the list. But soon those little black bugs would go from his garden to his workshop and beyond, launching a second career every bit as demanding as the first. Not surprisingly, the journey started with pollination.

"My fruit set was terrible," Brian told me, and explained how the forty pear and apple trees espaliered along his back fence had always bloomed vigorously, but never produced much. When he came across an agricultural bulletin on native pollinators, something clicked. "I suddenly realized those little black insects were bees," he said. Rushing outside, he found a small population of orchard mason bees moving among his fruit trees and flowering shrubs. Up close, their small black bodies shone with a bluish tint, and tufts of tawny hair covered their faces and fringed the base of each clear wing. He tracked their flight path to the garden shed and saw how the shingles overlapped to make perfect little nest holes. Each bee scurried in and out of its own crevice, slowly filling the gap with pollen and then closing it off with a carefully sculpted plug of mud. When Brian drilled holes into a piece of lumber, the bees filled those up too. He kept at it, and two years later had more mason bees (and more fruit) than he knew what to do with. On a whim, he decided to give the bees away as Christmas gifts.

FIGURE 3.1. Mason bees include over three hundred species in the genus *Osmia*. Here a male red mason bee (*O. bicornis*) peers from a nest hole. PHOTO BY ORANGAUROCHS VIA WIKIMEDIA COMMONS.

"Everybody loved it!" he said, showing me his original prototype—a small block of wood with a cute peaked roof and twelve empty nesting holes. Three additional holes, filled and capped by the bees, had been glued to the bottom. When Brian's friends and family hung these unusual gifts outside the following spring, dormant bees emerged from the mud, found the nearest sources of nectar and pollen, and promptly filled the empty holes with new nests. "It worked perfectly," he recalled. "Better than I expected."

For many people, the story might have ended there, with a memorable Christmas morning and a fun spring lesson in backyard pollination. But Brian brought a business sense to biology, and he smelled opportunity. When he took a carload of his bee houses to the regional garden show, he sold every single one of them. Soon he found himself supplying mason bees to individuals and retailers across North America. He attended bee classes, wrote bee books, and began lecturing about bees to garden clubs. He took in a business partner and hired out the construction of bee houses, bee chalets, and cardboard bee tubes as well as custom liners and refills stocked by an expanding network of enthusiastic bee growers. Nowadays, mason bees are sold everywhere from hardware stores to Amazon.com, but thirty years ago Brian was a pioneer. "All the information was out there," he assured me, and rattled off the names of experts and references that helped him along the way. Then he laughed and shook his head, adding, "But maybe it took an old insurance peddler to put it all together!"

In some ways, the success of Brian's bee business should come as no surprise. It capitalized on a lifestyle that has been flourishing in nature for over 120 million years. Orchard mason bees, like their sphecid wasp ancestors, are solitary creatures. Each female builds and provisions her own nest alone, without the cooperation of a hive, living out her adult life in a brief flurry of activity timed to coincide with spring-blooming flowers. Understanding and repackaging that strategy gave Brian more than just a thriving cottage industry. It revealed to him (and to his customers) a pattern of behaviors established long ago that continues, with small variations, to sustain the vast majority of the world's 20,000 bee species. We tend to admire evolution for its innovations—the transition from wasp to bee, or the inventions of honey and hive. But the process is also deeply conservative. Traits and habits that work well tend to stick around for a long time. Solitary bees embody that theme, one of evolution's lesser known, but equally important dictates: "If it ain't broke, don't fix it."

"Oh, she's going to lay an egg!" Brian exclaimed, as we watched a mason bee turn herself around and back into a nest hole. Dozens of others buzzed harmlessly about our heads, flying to and from a cluster of cardboard tubes and wooden bee blocks fixed to the back wall of his garden. Out of sight in her nest, the bee would be laying a single tiny egg onto a ball of "bee bread," the sticky mass of pollen and nectar she'd spent all day collecting. Her next foraging trip would be in search of mud to seal the egg into its chamber. Then she'd start all over again, repeating that sequence of pollen, nectar, egg, and mud until the entire nest hole was stuffed full. "They really are great masons," Brian said, and described how the bees mixed soil or clay to just the right consistency, shaping and polishing it with a coordinated motion of mandibles, forelegs, and abdomen. "I've taken nests apart and looked at them under a microscope," he went on with admiration. "The walls are perfectly smooth."

Now in his eighties and retired for a second time, Brian devotes most of his still-considerable energy to a new passion—making custom ukuleles. (Ever the businessman, he's sold over eighty of them to players and collectors around the world.) But he still maintains a population of bees in the garden, and I sensed no loss of enthusiasm as we sat there in the spring sunshine, watching them work. With his deep, steady voice and clear-eyed gaze, only Brian's shock of white hair suggested his age. As the afternoon progressed, it was obvious he still sipped regularly from that unrivaled wellspring of youth: curiosity. "Let's see if the mama bees can find their babies," he said at one point, rearranging two of the nest boxes. Within moments, several confused bees were walking around the empty shelf where their nests had been. While the distinctive whiff of a personal pheromone tells them which hole is theirs, they rely on visual landmarks and spatial cues to home in on the general location, another habit inherited from sphecid wasps. In time, these bees might sort out a change of inches, but larger alterations can make a nest site unrecognizable.

I couldn't help feeling sorry for those disoriented mamas, even though I knew they would never set eyes on their babies no matter where Brian moved the nest blocks. For solitary species like mason bees, the contract of parenthood ends with provisioning. Once the egg is walled in with its feast of bee bread, the mother moves on without a second thought, building and supplying new nest chambers in a mad, month-long frenzy. With good weather and plentiful flowers, a single mason bee can provision over thirty eggs before she's too worn out to continue. Once, I found an exhausted-looking female in our orchard and placed her on top of a new nesting block that I hoped to fill before the end of the season. It was in perfect habitat—a sunny spot surrounded by fruit trees, right next to a patch of muddy soil. She walked to the edge of the block and teetered for a moment, as if peering wearily at the rows of empty holes, and then toppled off dead into the grass.

The few weeks we see mason bees flying around make their lives seem brief and frenetic, but months of additional activity and intrigue, followed by a nice long rest, take place out of sight in the quiet darkness of their little adobe apartments. Already, eggs had begun hatching in the nest blocks on Brian's garden wall. If everything went according to plan, those tiny larvae would munch away on their bee bread all spring and summer, growing large enough to spin layered, silken cocoons. Like the better-known transformation that turns caterpillars into adult butterflies or moths, the life cycle of bees includes a complete metamorphosis. Inside those tough, waterproof cocoons, their bodies change from stout white grubs into the winged and fuzzy forms that we recognize as adult bees. Then they rest, dormant through the fall and winter until rising spring temperatures rouse them from their torpor. This process—in mason bees and thousands of other species—has been repeating itself for millions of years. It means that virtually anywhere you look, at any time of year, solitary bees are there—if not flying around, then tucked away in their hidden tunnels and crevices. For bee lovers, that is a cheerful thought, but it doesn't mean that life within a nest is always calm and copacetic.

"I'm glad you're here making me clean this up," Brian said, sounding embarrassed. "I've really let things go this year." To me, the garden seemed alive with bees, but Brian shook his head. "Just look at all the ones that didn't make it," he said, and began pulling out cardboard nest tubes that still had their mud caps intact. This far into the season, all the healthy, mature bees had already chewed their way out—those were the ones humming around our heads. Nests that lacked an obvious exit tunnel through the cap were failures, filled with bees that had fallen victim to mites, fungal infections, or worse.

"There," Brian said, and pointed out something different—a small, perfectly round hole that pierced the *side* of one of the cardboard tubes. Somebody hadn't left by the front door. "Are you familiar with *Monodontomerus*?" he asked, and rummaged through the discards for a minute longer. Then he reached over and dropped a tiny speck of metallic blue into my palm. Through a hand lens the insect sprang into sharp relief—a perfect wasp, smaller than a rice grain, its every surface shimmering with iridescence. I tilted it back and forth, watching the color shift from blue to green and gold in the sunlight. It looked like some kind of jeweler's fancy, the Fabergé version of an insect. But to mason bees, this little bauble, and others like it, was a deadly threat.

"They show up late in the season," Brian said over his shoulder, as he continued tidying up the nest blocks. For wasps in the genus *Monodontomerus*—or *Monos*, as they're known for short—there was no point arriving early. In fact, *Mono* females key in on the scent of cocoons and accumulated droppings, sure signs that the young bees within a nest have grown large and fat. What happens next is the kind of gruesome twist that shows up in horror movies. After sniffing out a promising nest, the *Mono* female inserts her long, needle-like ovipositor through the mud barrier (and in some cases even through surrounding wood) to pierce the cocoon; then she lays her eggs on the young bee. They hatch immediately and begin devouring their host alive, effectively transforming the mason bee nest into a wasp nest. Once satiated, the baby *Monos* use the cocoon just as the bee would have—as a sheltered place in which to rest and metamorphose before chewing their way out to freedom.

Brian's wasp infestation reminded me of a comment that Michael Engel had made. "Parasitism is the real story of the Hymenoptera," he'd told me, referring to the entire taxonomic order that included bees, wasps, and ants. The habit evolved early and often, he explained, and remains a dominant lifestyle in the group, particularly for wasps. When the larvae consume or otherwise destroy their hosts, as *Monos* do, entomologists call them "parasitoids," and nearly all bees have at least one of these species to contend with. Four different kinds of *Monos*, for example, attack the mason bees living in Brian's garden, as well as at least one *Chrysura* wasp and a parasitoid fly. (Though it's probably no consolation to their hosts, many parasitoids fall prey to other parasitoids, adding yet another layer of creepy exploitation to the life within a nest.) As if that weren't enough, bees also face treachery from within their own ranks.

"Maybe we'll see a cuckoo bee," Brian said, peering up at the mass of insects buzzing above us. With the nest blocks now in good order, we'd settled on the wooden rim of a nearby garden bed to watch. Seen from below, the bees showed off one of their most distinctive and charming traits. Masons belong to a large family that also includes leafcutter bees, who line their nests with scraps of greenery, and wool-carders, who use a felt of downy plant fibers. But while construction methods may vary, all members of the family carry pollen in the same location: on their bellies. The result makes each mama bee look like she's wearing a tiny, brightly colored apron—sometimes yellow, sometimes orange, pink, red, or even purple, depending on what type of flower she's been visiting. This cheerful habit sets them apart from nearly all other bees, whose pollen loads look more like stockings pulled up high on their back legs. To find a cuckoo, however, Brian and I needed to spot a bee with no pollen at all.

The word "cuckoo" comes directly from nature: it's a medieval French term meant to mimic the two-note song of the bird it describes. Owners of certain clocks know that maddening phrase well, but cuckoos are also famous for laying their eggs in other birds' nests. This ploy allows them to avoid the demands of child rearing, since the host bird will raise their chick as one of its own. Cuckoo bees do much the same thing, but since most bees are like masons, and don't tend their offspring directly, what a cuckoo is really avoiding is the painstaking collection of pollen and nectar. Instead of spending long hours searching for suitable flowers, it simply darts into a nest cell and lays its egg when the resident bee is away. If this trickery goes unnoticed (and most cuckoo eggs are very well camouflaged), the host bee completes her work without suspecting a thing, sealing the foreign egg inside with her own. As soon as the eggs hatch, the intruding larva murders the rightful tenant with a pair of specially adapted sickle-shaped mandibles, and then settles in to feast on its cache of purloined bee bread. Biologists call such creatures "klepto-parasites," a Greek phrase for anyone that makes a living by stealing food from others. It's an apt title for many college roommates, but also describes a surprising number of bees.

"At least 20 percent . . . probably more," Michael Engel esti-mated, when I asked him how many of the world's bee species were parasitic. Like the solitary habit, kleptoparasitism ranks among the unsung success stories of bee evolution, measured, in this case, by the sheer number of times it has cropped up. Stealing trumps gath-ering for thousands of species in at least four of the seven recognized bee families, though it's difficult to get an exact count. Because the freeloaders don't need to collect pollen, they often lack hair and other bee-like characteristics, making them devilishly hard to iden-tify. Many appear wasp-like, and most are furtive and inconspicuous, useful traits for a lifestyle that relies on deception. But because they often specialize on one or a few closely related species, cuckoo bees have proliferated right alongside their hosts. New bee species beget new cuckoos ad infinitum, adding a fascinating layer of diversity and complexity to the evolutionary story of bees.

Brian Griffin and I never spotted a cuckoo among his masons. The bees hovering and darting above us all wore golden aprons of pollen, or sometimes appeared clasping smooth balls of mud in their jaws. But if we'd watched for a full season instead of just one after-noon, the cuckoos would surely have made an appearance—lured by the promise of bee bread and a dry home for their offspring. To-gether with Monos and other parasites, they transform the simple-looking nests of solitary bees into remarkably competitive and dangerous places. Masons counter these threats by guarding their nests whenever they're not actively foraging. (Peer inside and you'll often see the furry face of the mother bee glaring right back at you.) They also seal their doorways with an extra-thick mud plug that, like the entrance to a pharaoh's tomb, leads first to an empty ante-chamber before the real nest cells begin. And, just like the ancient Egyptians, masons hide their most valuable treasure at the farthest end of the tunnel.

"I learned that a six-inch tube works the best," Brian told me when we toured his workshop, inspecting an array of designs he'd experimented with over the years. "Anything shorter and you get too many males."

This odd statement revealed something fundamental about bee biology: males are expendable. Like ants, wasps, and various other insects, mother bees can predetermine the sex of their offspring: fertilized eggs produce females, while the unfertilized ones grow up to be male. They control this switch by doling out sperm stored from their mating flights in a special pouch near the base of their ovaries. This system allows mason bees to play the odds, concentrating their precious female progeny beyond a certain depth in the nest hole, where any parasite (or, for that matter, a hungry woodpecker) would have to breach all the intervening cells to reach them. Brian had a glass-fronted display nest that showed the situation perfectly—the deeper female cells looked pampered, stuffed with bee bread and half again as large as the scantily provisioned male chambers near the entrance. For anyone building bee nests, this system puts helpful parameters on the appropriate depth of the holes. For male bees, it offers only a shrug of cold logic: so long as enough of them survive to breed, the population can afford to lose the rest. As consolation, those males who do make it through until springtime lead a relatively easy life. By position and design they emerge first, with any laggards quickly nipped and nudged into action by the bees coming from behind. Once outside, they loiter near the nest site, maybe tussle a bit for position, and then pounce upon and mate with any

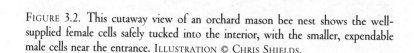

FIGURE 3.2. This cutaway view of an orchard mason bee nest shows the well-supplied female cells safely tucked into the interior, with the smaller, expendable male cells near the entrance. ILLUSTRATION © CHRIS SHIELDS.

and all females they can find—usually the very instant those females crawl out of their holes. With that task taken care of, males can then dither away the few days of life remaining to them while the mama bees get on with the vital work of providing for the next generation.

Nest design and other habits vary, but the same basic events that make up an orchard mason's life play out in a similar way for nearly all the world's solitary bees. Some species dig holes in hard-packed soil or sand; others use hollow twigs, pinecones, or the grooves in tree bark. I've found bees nesting in compost heaps, sidewalk cracks, firewood, rock piles, rolled-up umbrellas, and the notches in a block of surfboard wax. There is an Indonesian bee that nests inside active termite mounds, and one in Iran that glues together elaborate vases of pink and purple flower petals. Over two dozen European and African species nest exclusively in the abandoned shells of snails, and at least two North American varieties make their homes in desiccated cow pies. But regardless of where they do it, all of these bees follow the same ancient cycle of emergence, mating, nest construction, provisioning, and egg-laying. And, like masons, they all play host to a range of cuckoos and other parasites, meaning that any given nest might produce multiple species of bees as well as wasps, flies, and even beetles. While the solitary lifestyle is clearly successful, it is also perilous. The constant threats of parasitism and predation may help explain the evolution of another defining bee characteristic: not all of them choose to live alone.

FIGURE 3.3. Nesting in aggregations may offer solitary bees some of the advantages enjoyed by animals living in herds: lowered predation risk, group defense, and the intriguing potentials of life in a new evolutionary context. Elbridge Brooks, *Animals in Action* (1901). WIKIMEDIA COMMONS.

"Here's a question I've had for a long time," Brian Griffin said to me, near the end of our afternoon together. "If these bees are solitary, then why are they so gregarious?" He showed me several cracks in the rock wall near his front door, where a few bees were indeed nesting off on their own. But the vast majority of his masons always clustered in one area, no matter how he arranged the nest blocks. "They seem to want to be together," he mused. "Why is that?"

For some bees, crowding is an unavoidable consequence of limited habitat; cliffs, patches of exposed soil, or suitable holes in bark, twigs, and wood are often scarce commodities. But at least part of the answer lies in the old biological saw about "safety in numbers." If you are a single zebra, for example, and you walk past a hungry lion hidden in the grass, you are a dead zebra. If you're with a whole herd, on the other hand, your chances of survival improve dramatically. Herding lessens the risk to any particular zebra through sheer probability. But it also offers opportunities for group defense, and for the evolution of subtleties like stripes (which some experts believe can visually confuse predators at close range). For solitary bees, the logic is similar. Nesting in aggregations helps diffuse the risk posed by cuckoos and other parasites. But where it really gets interesting are the subtleties. When solitary individuals cluster together, generation after generation, the simple fact of their proximity opens the door to new behaviors. Some species, like orchard masons, remain steadfast in their solitude—one female per nest. But others have experimented with cooperation, from occasional nest sharing to collective provisioning, brood tending, and defense. On at least four separate occasions, this path has brought about levels of complexity that experts refer to as "eusocial," or "truly" social. We recognize eusociality from the highly organized, hive-building habits of the bees we know best—honeybees. But if one of the field's most prominent thinkers is correct, our familiarity with this lifestyle runs much deeper.

In his 2012 book, *The Social Conquest of Earth*, Harvard biologist E. O. Wilson laid out the crucial prerequisites for eusociality: multiple generations living together, the division of labor, and the presence of altruism. The rare creatures to attain this combination, including ants (his specialty) and termites, as well as certain wasps and bees, have often enjoyed extraordinary success in nature. To this short list, Wilson proposed an unorthodox addition: people. As he put it in an interview, one of the handful of species—and the only large animal—to finally meet all the eusocial criteria "happened to be a big primate in Africa."

Not surprisingly, Wilson drew immediate criticism for lumping humanity in with a group of organisms dominated by insects, a few shrimp, and the naked mole rat. But he was hardly the first person to point out similarities between human societies and the habits of creatures like honeybees. Scholars have held up the hive as a model for humanity since at least the time of Virgil, who wrote of bees, "They alone share the care of their young and live united in one house, and lead lives subject to the majesty of law." A good part of the controversy surrounding Wilson's claim centered on his theory for *how* eusociality evolved—not just through the relative survival of individuals, the traditional view, but by natural selection acting on entire groups. This way of thinking offers an intuitive explanation for altruism. Self-sacrificing traits that seem at odds with "survival of the fittest" (such as reckless valor in battle, or giving up breeding opportunities) can still persist, and even flourish, if they benefit the group as a whole. But Wilson's model flies in the face of decades of work based on a mathematical formula for degrees of relatedness (altruism persists in the gene pool only if it benefits enough close kin to outweigh its considerable personal costs). The matter remains far from settled, but there is one point on which all the experts can agree. If you want to study social evolution in action, there is no better place to look than the lives of bees.

For other well-known groups, the transition to eusocial living took place only once, far in the distant past, and it produced descendants that all lived more or less the same that way. Termites evolved from solitary, cockroach-like ancestors over 140 million years ago, and ants came from solitary wasps not long after. Together, they now account for an estimated 25,000 highly social species. If we accept Wilson's premise, the primate genus *Homo* crossed the eusocial threshold 3 million years ago and never looked back (even if some of its members do spend a lot of their time alone, sitting in shacks, writing books). For bees and certain wasps, however, the story is quite different. A lifetime of study taught the great entomologist Charles Michener to be cautious on this topic. "Clearly, there is no ready answer," he wrote, when trying to sum up the number of times eusociality had evolved in bees. Honeybees and their relations were obviously social, but other groups appeared to have invented the habit and then done away with it, while some teetered on the verge, difficult to classify. In fact, degrees of sociality could change within a single population, or even for a single bee, over the course of a season. "It is the wrong question," Michener concluded, implying that the most interesting matter at hand was more fundamental: Why do bees display such a dizzying spectrum of social behaviors in the first place?

If I'd started writing this book a few years sooner, I could have asked Michener that question directly. He was famously approachable—and still busy with research—right up until his death in 2015, at age ninety-seven. Instead, I did what most people curious about bees wind up doing again and again. It's a bit like the parlor game "Six Degrees of Kevin Bacon," where movie buffs try to connect anyone in Hollywood to a Kevin Bacon film in six steps or less. In the world of bees, it doesn't take nearly that long to get to Charles Michener. I had already talked with two of his graduate students—Michener served on Jerry Rozen's doctoral committee in the 1950s, and on Michael Engel's in the 1990s. Now I moved one step further afield and visited one of his student's students, a leading entomologist who had been thinking about the evolution of sociality since long before he knew anything about insects.

"My first degrees were in history and linguistics," Seán Brady told me, and described his early fascination with social development in the human realm. He had only turned his attention to insects after reading a book about ants, when he realized how little was known about their evolution and the origins of their own complex sociality. "I thought, 'I can do better than that!'" he recalled. It was a career decision that took him quickly from ants to bees and a postdoctoral position with Michener protégé Bryan Danforth at Cornell University. Now working as a department head at the Smithsonian Museum of Natural History in Washington, DC, Seán has arrived, almost inevitably, at a group of sweat bees whose peculiar social habits were one of Charles Michener's abiding passions.

"I wonder if Mich collected any of these," he said, as we peered into a box full of tiny, black bees. We were standing between rows of tall white cabinets that moved on tracks set into the floor. The system limited access to one narrow aisle at a time, but doubled the capacity of the room, a space-saving necessity when trying to organize and store over 35 million specimens. But while they belonged to one of the world's largest insect collections, the bees in question were so small they couldn't be stuck with pins. Instead, their bodies had been painstakingly glued to the sides of the pins, lined up in row after indistinguishable row. Even Michener had admitted that their appearance was "morphologically monotonous." What set these bees apart was their lifestyle.

"We know that climate influences their sociality," Seán said, going on to explain how the particular species we were looking at was solitary in the colder parts of its range and eusocial in the south, where warm weather extended the nesting season and allowed mothers and daughters to interact. Later, he showed me pictures of a tropical species where the mother produced both small daughters, whom she could boss around as nest helpers, and larger, well-fed daughters who dispersed and bred. In other cases, a mother might raise a social, all-female brood early in the season and then perish, leaving it to her daughters to produce males, breed, and disperse to establish new nests. Though no sweat bees have quite achieved the elaborate hive societies made famous by honeybees, hundreds of species display aspects of altruism and overlapping generations, the hallmarks of eusociality. Their evolution helps explain why bees in general have developed a wider variety of social behaviors, and done so more often, than all other insects combined.

"We know that something about their nesting behavior is involved," Seán told me, when I asked him why his sweat bees seemed so predisposed to sociality. "They tend to have specialized nesting sites that are patchy and limited," he explained, which forced them to live together. "So they just kind of learn to get along." But while that kind of communal living is important, it doesn't necessarily lead to sociality—Brian Griffin's mason bees, after all, live side by side in their nest blocks but rarely interact. Perhaps the most critical factor is not what happens between unrelated females, but what happens to their daughters. What impulse would drive them, at least on occasion, to stay and help tend the nest, rather than dispersing to reproduce? Seán called the origin of that behavior "elusive," but pointed to a breeding system, shared with wasps and ants, that made it at least somewhat more likely. Because males come from unfertilized eggs, they contribute reduced genetic variability to any given brood, making all the sisters in a nest particularly close relatives. Genetically, that translates into a bigger payoff for altruism—helping your mother or sisters provide for the next generation will pass on a lot of your genes, too, even if you forfeit your own breeding opportunities to do so.

"Sociality seems to flicker on and off in these bees," Seán said later, noting how the behavior had evolved on two or three separate occasions, 20 million years ago, and spread through two of the largest genera in the family. But then various descendants had lost it at least twelve times, reverting back to a solitary existence. That situation differs markedly from that of other insects, such as ants and termites, where eusocial behavior evolved once and became fixed. In one of Seán's major papers on the subject, he and his coauthors suggested that sweat bees were simply new to the social game, so their habits were still in flux (20 million years isn't considered all that long in evolutionary terms). "But on the other hand, it could be something we don't know about yet," he mused, and his eyes lit up. Watching Seán think, it was obvious he had a true scientist's love of the counterargument, like a lawyer who can't help picking apart his or her own case. "Maybe something will show up in the genetic data, some quirk that makes them socially flexible."

We had moved from the insect collection to his office, an unadorned room with a window looking out onto a blank wall. There were signs everywhere of research in progress—boxes of specimens, racks full of vials, and stacks of papers on the desktop, the table, the chairs. Shelves lining the walls contained books, more boxes, and, I was pleased to see, two hairdryers—an essential tool for fluffing up the fuzz on wet or otherwise disheveled bee specimens. Seán himself looked a little bit frazzled, and he rubbed his eyes wearily at several points during our conversation. The administrative burden of chairing a large entomology department took up more and more of his time, and had recently forced him to cancel a much-anticipated collecting trip to South Africa. But when I asked him what his research group was working on, his face brightened again and he told me about an ambitious genetics project analyzing data from a wide swath of bees and wasps in the collection. The resulting family tree, time-stamped with fossil evidence, would help to establish how and when various bees and their social habits developed. "It's like being a naturalist back in the nineteenth century," he said, describing the potential of the new genetic tools. "We're on a big fishing expedition at this point."

I left Seán's office better informed, but still puzzled about the social complexity of bees. Perhaps Charles Michener was right. The best answer is to keep questioning, which is exactly what Seán Brady and other specialists are doing. Maybe with genetics and a few more fossils, the paths that bees travel to (and from) sociality will become clear. For now, it must be enough to know that whenever solitary bees nest close together, the stage is set for interaction. Often nothing happens, but sometimes they start to cooperate, and once in a while, a daughter stays home to help her mother. And if those first tentative steps are successful and lead to more, the results can be spectacular.

On the museum's busy second floor I made my way past throngs of schoolchildren and a long line of people waiting to enter a live butterfly enclosure. Finally, built into the wall of a corner room called The Insect Zoo, I found a small display hive of what many people consider the most socially advanced creature on earth, the honeybee. Hundreds of scientific careers and countless books and papers have been devoted to describing honeybee habits—how a single breeding queen surrounds herself with daughters divided into organized, task-based castes that forage, defend, clean, make honey, and tend a growing brood. It was December, and the bees had apparently been moved to other quarters. There wasn't much to see—just a few dead workers and some desiccated honeycomb. But on a previous summertime visit I'd watched bees industriously flying in and out, making their way through a long Plexiglas tube that connected to the world outdoors, where the landscaping was in full bloom all around the three-hundred-acre National Mall. With such an abundance of pollen and nectar on hand, a single hive could easily grow to more than fifty thousand individuals, quite a testament to the eusocial habit. Eleven species of honeybees as well as hundreds of closely related stingless bees live variations of that lifestyle across southern Europe, Asia, Africa, Australia, and throughout the tropics. Wherever they occur, domesticated or wild, these highly social species are often the most common bees in the landscape, critical as pollinators and also as producers of honey (sustenance for thieving birds and mammals as well as the hive). E. O. Wilson describes the nests or hives of such creatures as a single teeming extension of the life of the queen, cooperating in what social biologists have marvelously labeled a "superorganism."

With that kind of incentive, it's not surprising that the path to sociality has arisen more than once. Evolution is like that, a relentless process of reinvention that often arrives at the same solutions again and again in different situations. Bees live in a huge diversity of habitats, where the various levels of solitary, communal, and social living all have something to offer. Over evolutionary time, groups of bees have shifted among these lifestyles to best take advantage of their particular situation. This all made perfect sense, but it left me with a nagging, and in some ways more fundamental, question: If bees themselves are so successful, with thousands of species playing vital roles in ecosystems around the world, then why hadn't the pollen-eating habit evolved again? Why had only one group, among all the carnivorous wasps buzzing around over millions of years, made that critical transition to a vegetarian lifestyle? I decided to put this question to Michael Engel, and his answer was immediate.

"*Krombeinictus!*" he said with enthusiasm, then directed me to a paper about a small sphecid wasp living a decidedly bee-like lifestyle in the hills of Sri Lanka. The reference was two decades old and rarely cited, but with luck and persistence I managed to track down one of its coauthors. She told me a tale of intrepid scientific discovery, where the prize was a new species that behaved like no other known sphecid. Inadvertently, her story also revealed something vital about the evolution of bees and the flowers that sustain them.

Bees and Flowers

Of course you know that bees could not exist without flowers; but do you know that many flowers could not exist without bees?

—Rev. Charles Fitzgerald Gambier Jenyns,
A Book About Bees (1888)

CHAPTER FOUR

A Special Relationship

*The botanist should make interest with the bees if he would
know when the flowers open and when they close.*

—Henry David Thoreau,
Journal entry (1852)

The monsoons arrived late to Gilimale, Sri Lanka, during the
summer of 1993, transforming its simple, backcountry roads
into rivers of impassable mud. "If you wanted to go anywhere during
the rains," Beth Norden recalled, "you either rode an elephant or
walked."

The unexpected weather shortened her field season to a few fran-
tic days spent cutting twigs from branches and stuffing them into
old shampoo bottles for analysis later. It rained again when she returned in 1997 on a Fulbright Scholarship, but by then she knew
she was onto something. "When we started to figure out what was
happening, we said, 'Nobody's going believe us—they're going to
think we're making it up!'"

The twigs Beth took home to her lab at the Smithsonian came from a small tree in the pea family known for its friendliness to ants. It produces hollow nesting places near its branch tips for the ants to live in and also gives them abundant nectar to drink. In exchange, the ants vigorously defend the tree against anything that tries to eat its foliage. (Cleverly, the tree's nectar oozes not just from its flowers, but also from glands on buds and young leaves, attracting its ant defenders to the tenderest places most vulnerable to attack.) When Beth began to open up the hollow twigs, she found the expected ants in abundance, as well as spiders, springtails, bees, parasitic flies, and, very rarely, the nest of a small black, yellow, and reddish sphecid wasp. That's when she noticed something unusual.

"The wasp's larvae looked yellow, like they'd been eating pollen," she told me, and explained how bee grubs often picked up a hint of color from their flowery diet. But her colleague and mentor on the project, the late Karl Krombein, was doubtful. Over decades of research, he'd built up a reputation in the wasp community that was something akin to Charles Michener's status in the world of bees. He had discovered and described scores of new species, many of them in Sri Lanka. But he'd never seen anything like this. The nests lacked any trace of arthropod remains—whatever those little grubs were eating, it wasn't the typical sphecid diet of paralyzed flies and spiders. Then they found another clue: a female with pollen grains stuck to the hairs around her mouthparts. Finally, a microscopic analysis of the larvae's droppings revealed digested pollen in abundance. That settled it—like that elusive proto-bee in the Cretaceous, Beth and Karl's new sphecid was a hunting wasp that had given up hunting.

"We were just at the right place at the right time," Beth told me modestly, when I reached her by phone. Long retired now, she seemed pleased to reminisce about the species that still bears her and Karl's surnames, *Krombeinictus nordenae*. "I think it specialized on the tree for nesting first," she mused, "and then there would have been all sorts of reasons to switch to pollen." Once established in their twig ends, the wasps would have found themselves surrounded by the same nectar sources that fed the ants, plus, during the flowering season, a rich source of pollen. Abandoning the hunt made it possible for an individual wasp to complete its entire life cycle within the crown of a single tree, and Beth doubts they're even found on any other species. That might explain why Karl hadn't seen one on any of his fourteen previous trips to Sri Lanka, and why, to Beth's knowledge, nobody has collected one since. (Even when you're looking for them, they're hard to find—splitting apart thousands of hollow twigs yielded Beth and Karl only nine adult wasps, so few they couldn't even spare one for a dissection.)

The story of Beth's wasp begs an obvious comparison and question. For the sphecid ancestor to bees, switching to vegetarianism gave rise to a lineage of incredible abundance, diversity, and prominence. So why should *Krombeinictus* make the very same dietary change, yet remain so utterly scarce? It could be that its ancestors started on pollen only recently, and great things lie ahead. Certainly, *Krombeinictus* displays many of the traits and behaviors associated with early bees—small, solitary, and specialized on particular flowers. (Intriguingly, *Krombeinictus* also shows signs of early social evolution. Mothers display a high degree of maternal care and rear each larva to adulthood in an open nest, providing opportunities for generations to overlap and cooperate.) But it's just as possible that pollen-eating wasps crop up now and then without making much of an evolutionary splash. "I have no doubt there are others out there doing the same thing," Beth said. "We just don't know about them."

In fact, there is another group of vegetarian wasps only marginally less obscure than Beth's. The family Vespidae is best known for its stinging hornets and yellow jackets, but it also includes a group of pollen eaters that evolved around the same time that bees did, and have been plugging away quietly ever since. These "pollen wasps" now number several hundred species worldwide, but have never achieved widespread ecological prominence. Few people have ever seen one, and fewer still would know if they had. (Even Michael Engel gave them only two sentences in his insect evolution book.) A vegetarian diet alone, then, cannot explain the rise of bees. Their success also stems from how that diet has changed them, and how they, in turn, have changed the plants that provide it.

Winston Churchill coined the phrase "special relationship" during a speech he gave in the spring of 1946, a particularly memorable commentary on world affairs that also introduced the term Iron Curtain. Churchill was referring to the shared cultural, economic, and military interests that made the United Kingdom and the United States particularly close, an exceptional alliance that stood above all other diplomatic ties of either country. Plants and animals can have special relationships, too, ecological associations of extraordinary consequence. Over time, those interactions may lead to coevolution, changes in the inherited traits of the various partners in the dance. Textbooks often portray this process as a pairing, a quid pro quo, but it's almost always more complicated, involving multiple species and environmental effects that vary widely across time and geography. Ecologist John Thompson introduced me to a wonderfully vivid term for these interactions, "coevolutionary vortices," like whirlpools that form and drift along within the larger stream of evolution itself. But in spite of its complexity, people usually identify coevolution by the relatively straightforward signs it produces in the major players—a faster antelope begets a faster cheetah, and so on. And for bees, the most obvious result of their long dance with flowers boils down to one thing: fuzz.

In children's poetry, the fuzziness of bees always comes up because it rhymes so well with another one of their telltale traits, their buzz. But even scientists often rely on hairs to identify and describe

the bees they study. One glance at a bee's plush coat can be enough
to differentiate it from a wasp, particularly under magnification,
where the unique quality of bee hair becomes obvious. For wasps,
the sparse hairs scattered across their smooth bodies look simple,
like short, pointy threads. Bee bodies, on the other hand, boast a
mixture of hairs—some simple, but some as branched and downy

FIGURE 4.1. A bumblebee forages on a coneflower (top), its
body festooned with pollen. In the scanning electron image
(above), individual pollen grains cling to a bee's distinctively
branched hairs. TOP PHOTO © RICHARD ENFIELD; BOTTOM
PHOTO © UNIVERSITY OF BATH, UK.

as feathers. And just as the feathers in a duster quickly collect small particles from shelves or lampshades, so do the hairs of bees collect pollen. Their complex surfaces give pollen grains a lattice of nooks and crannies to cling to, vastly increasing the efficiency of bees as pollinators. Spend any time watching flowers and you can see this in action—bees festooned with pollen will often forage alongside nectar-sipping wasps, whose smooth bodies remain pristine. For a more precise test of this idea, however, there is a simple experiment I recommend that involves nothing more than wheat flour, an accurate scale, and a pair of appropriate insects.

Natural history museums like the Smithsonian preserve their collections in rows of hermetically sealed cabinets designed to keep out moisture, pests, fungi, or anything else that might threaten the pinned and labeled specimens stored within. I use an ice chest. Though designed for chilling snacks and beer, any midsized cooler with a tight lid (and a few mothballs) also works perfectly well as a bug-saver. The experiment I had in mind required only two specimens—a sphecid of the sand wasp variety I'd been watching at the gravel pit, and a bumblebee of similar size. Lying side by side on my office workbench, the two insects looked similar, and it was obvious the bee had inherited a lot from its wasp ancestors—it had the same basic body shape and a similar set of paired, delicate wings. But where the wasp looked long and sleek, with only a scattering of spine-like hairs across its back and legs, the bee appeared as stout and heavily furred as a small mammal in winter. (Psychologically, this may be another aspect of the human affinity for bees: some of them, at least, look like animals we'd like to pet.) After carefully weighing each insect on the scale, I covered the bottom of a Petri dish with flour and dumped them both in.

Dredging dead bugs in pulverized grain may sound like a crude substitute for pollination, but my results were surprisingly informative. The flour behaved beautifully, separating into tiny white clumps that stuck to insect hairs just like the real thing. Orchardists know all about this, and they often mix flour with pollen in ratios as high as nine to one when artificially fertilizing date palms, pistachios, or other high-maintenance trees. (Like adding water to soup to feed more people, this technique can stretch a small amount of pollen to reach a larger number of trees.) I lifted the bee out first. Flour covered its body like fake snow on a shopping mall Christmas tree, perfectly edging each leg and coating every tuft of exposed fuzz from stem to stern. I tapped the bee lightly and even blew a gentle wind across it, but most of the flour stayed put. On the scale, the bee's weight had increased by 28.5 percent, which would equate to something like a fifty-pound (twenty-three kilogram) backpack for an average-sized person. That's a pretty good haul for a stiff, inanimate specimen, and it's not surprising that live individuals do even better—wild bumblebees have been caught carrying pollen loads in excess of half their body weight. Turning my attention to the wasp, I saw that it, too, carried a dusting of flour. But if the bee's load was like heavy snowfall, then this was the sort of faint trace that would surely disappoint skiers, snowboarders, or any child hoping for a day off from school. A few flecks of white clung to the spiky hairs on its abdomen and legs, but most of the wasp's body looked perfectly clean. My scale, accurate to the hundredth of a gram, registered no discernable increase in weight.

The evolution of branched hair gave bees an advantage measured in one of nature's most vital statistics—more food for their babies. But it also scattered that pollen across the surfaces of their bodies, greatly increasing the likelihood that at least some of it would brush off onto other flowers. The inherent sloppiness of a fuzzy body goes a long way toward explaining why bees flourished while other vegetarian wasps have never amounted to much. Beth Norden did find pollen on the hairs around her wasp's mouth, but she suspects they mostly just swallow the stuff and then spit it up again back at the nest, which is exactly what the vespid pollen wasps do. This habit may keep their larvae fed, but it eliminates the need for external traits like branched fuzz, severely limiting their potential as pollinators. After all, from a plant's perspective, what's the point of attracting smooth-bodied visitors that uselessly carry pollen around *inside* their bodies? And without a serious commitment on the botanical side of the equation, wasps have only occasionally become intimate players in the pollination process. Because it is that floral investment—what plants have done for bees—that makes the relationship coevolutionary. With both sides constantly adapting to the costs and rewards of pollen transfer, bees and their flowery hosts exist in a vortex capable of spinning off adaptations, and even new species, at a remarkable rate. In the mid-nineteenth century, the results of this relationship led to one of science's most notorious conundrums.

While bees appear rarely in the fossil record, flowering plants occur in relative abundance, appearing in Late Cretaceous deposits with such sudden diversity that they challenged Charles Darwin's notion of slow, incremental evolution. In a letter to the botanist Joseph Hooker, Darwin famously called the rise of flowering plants an "abominable mystery." Less famously, his letter went on to reference French scientist Gaston de Saporta's suggestion that "there was an astonishingly rapid development of the high plants, as soon [as] flower-frequenting insects were developed & favoured intercrossing." Darwin corresponded with Saporta for years and agreed that, if the plants had indeed evolved rapidly (a big "if," in Darwin's view), then Saporta's insect theory was the best explanation. In the end, both men were proven partially correct. Flowering plants did evolve before the Cretaceous, as Darwin suspected, and plodded along slowly for millions of years before their sudden proliferation. But Saporta gets credit for the more sweeping insight: how coevolution with insects, particularly bees, helped flowering plants dominate the

FIGURE 4.2. Charles Darwin corresponded for years with French naturalist Gaston de Saporta, whose beard might not have been as long, but who was the first scientist to suggest that coevolution with insects spurred the rapid evolution of flowering plants. WIKIMEDIA COMMONS.

earth's terrestrial flora, at the same time giving them many of their most recognizable features. Without that interplay, things would look and smell quite differently in our gardens, parks, hedgerows, and meadows.

When Henry Wadsworth Longfellow called flowers "so blue and golden," he probably wasn't thinking about the visual receptors in bees' eyes, but the prevalence of those shades in his much-pondered bouquet was not a coincidence. They fall right in the middle of a bee's visual spectrum, and flowers adopt them specifically in a bid to woo bees as pollinators. The evolution of petal color often tracks closely with a plant's strategy for getting its flowers fertilized—all the hues from mustard to cornflower would be exceedingly scarce, and might not exist at all, if there had been no need to advertise for the services of bees. Purple would also be rare, though there would still be a few perky splashes of red to tempt nectar-loving birds.

Scent is also a common bee-related trait, and Walt Whitman made a fine, if unintended, biological observation when he pined for a beautiful flower garden "odorous at sunrise." Many floral fragrances do indeed surge during the morning hours, just as temperatures rise and hungry bees become active, seeking out flowers that have filled with nectar overnight. For plants, it is a perfect pollination opportunity and a ripe moment to advertise. If there were no bees in the equation, Whitman might have timed his walk for a moonlit night, to get a good whiff of the cloying perfume given off by moth-pollinated flowers. Or he might never have considered an amble through a garden in the first place, since the majority of blossoms would reek of the musky terpenes and rotten flesh smells attractive to flies and wasps. (The fact that bees prefer odors we find worthy of poetry counts as one of nature's happier accidents.)

Beyond color and smell, the very shapes of many flowers can also be traced to bees. While round blossoms generally appeal to all sorts of pollen and nectar seekers (bees included), most of the more elaborate flowers evolved with specific visitors in mind. From an insect's perspective, round blooms can be approached from any angle or direction with the same result, a "come one, come all" display that often draws a crowd. If Claude Monet had included pollinators in his sunflower still lifes, he would have found himself busy adding in all manner of bees, as well as hover flies, bee flies, butterflies, wasps, and beetles. Flowers that diverge from round, however, can be more choosy in whom they invite, and where they deposit their pollen. The wide banners of a pea flower, or the lipped tube of a snapdragon, are what botanists call *zygomorphic*, from the Greek word for a yoke used to tether two oxen. Like any two-beast harness, they display bilateral symmetry, a concept that is also familiar from the sight of our own faces—draw a line down the center, from top to bottom, and one half is a mirror image of the other. For flowers, this design creates clearly defined sides as well as a distinct sense of up and down, requiring their visitors to enter in a specific way. Once that feat is accomplished, flower parts can develop all sorts of adaptations for dabbing pollen in particular places on insects of a particular shape and size. But plants can only afford such a focused approach if their pollen is likely to stick well to their intended targets, which makes bees by far the most common callers at zygomorphic blooms. Compared to his sunflowers, Monet would have found painting pollinators on his yellow irises a cinch, since bumblebees are virtually the only insects capable of getting the job done. With their deep tubes and upright orientation, irises force the bees to land on a designated platform and pass beneath a broad, pollen-laden stamen placed, as one expert pleasingly described it, "to fit exactly the dorsal surface of

FIGURE 4.3. If Claude Monet had painted still lifes with their pollinators, the round sunflower inflorescences on the left would have required a range of insects, from various bees to flies, wasps, butterflies, and beetles. For the more specialized irises on the right, he would have needed only bumblebees. WIKIMEDIA COMMONS.

the humble-bee." The female parts are there too, ensuring that the bee will deposit its pollen load in just the right location on the next iris it comes to.

When distinctive floral motifs evolve again and again to attract a particular suite of pollinators, botanists call them "pollination syndromes." These can include traits as general as flower size and color scheme, or as specific as the chemistry of an odor or the types of sugars that sweeten the nectar. Hummingbirds favor red, tube-like flowers rich in sucrose, for example, a syndrome that has arisen separately in plant families as different as honeysuckles, mints, figworts, buttercups, and mistletoes. Other floral patterns apply to everything from bats (pale, night-blooming, exposed) and butterflies (large, colorful, fragrant) to marsupials (brushy, robust, drab). While there are always exceptions, and many generalist flowers that attract multiple groups, pollination syndromes can be very useful for predicting interactions between plants and animals. Knowing the flower habits of moths, for example, allowed Charles Darwin to intuit the existence of a particularly long-tongued species from Madagascar four decades before its discovery. He never traveled to the island, but when someone sent him a Madagascan orchid that was fragrant and white, with a foot-long, nectar-filled spur, Darwin recognized at a glance that it could hardly have been pollinated by anything else. He immediately described the flower in a letter to Joseph Hooker, adding: "What a proboscis the moth that sucks it must have!"

As the most abundant and diverse of all pollinators, bees visit the greatest variety of flowers, attracted to a broad range of shapes and colors and often able to sneak onto blossoms better suited to others. (Most bees can't perceive the color red, for instance, but they can still spot many hummingbird flowers by their shapes and by the tonal contrasts between the blossoms and surrounding leaves.) In fact, the traits attractive to bees are so wide-ranging it's impossible to define a single "bee syndrome." Remove bees from the system and flowers lose all sorts of the attractive characteristics we take for granted, a fact that should have been obvious to the marooned sailor whose story inspired Daniel Defoe's novel *Robinson Crusoe*.

When Alexander Selkirk demanded to be put ashore on the Juan Fernández Islands in 1704, he expected his other crewmates to join him in deserting their captain's leaky, worm-eaten ship. None did, so he found himself alone on a rocky island, isolated in the cold South Pacific over 400 miles (650 kilometers) off the coast of Chile. Selkirk kept no diary of his four-year ordeal, but he reportedly became so skilled at living off the land that he could run down the island's feral goats barefooted and barehanded. If he matched his hunting prowess with an equal knack for gathering, Selkirk must have known the flora well, and might have wondered why nearly every flower he encountered was small, round, and greenish white.

Like other remote archipelagos, the Juan Fernández group has been slowly colonized by vegetation arriving from the continents. But while it now supports over two hundred plant species in habitats ranging from grasslands to dense forest, the only known bee is a tiny, rare sweat bee thought to have arrived recently from coastal Chile. It doesn't yet play a significant role in pollination, which means that for the several million years since these volcanic crags first rose from the sea, any bee-dependent colonizers have either failed to establish themselves or have had to adjust their pollination to the available methods, primarily wind and birds. Amazingly, plants from as many as thirteen different genera have learned to do just that. Some have lengthened their flowers to better accommodate hummingbird bills, while others, that now rely on wind, betray their pollinator-dependent roots by continuing to produce copious nectar, a reward for bees that never come. The Juan Fernández flora established and developed in the absence of bees, and its monotonous green and white flowers hint at how a bee-free world might look. But at the same time, the ability of at least some new arrivals to change their

FIGURE 4.4. Illustrations of Robinson Crusoe often depict the castaway surrounded by lush tropical vegetation festooned with flowers. On the islands that inspired the story, however, most flowers are small and drab because of an almost complete lack of bees. Illustration by Alexander Frank Lydon, in Daniel Defoe, *The Life and Adventures of Robinson Crusoe* (1865). WIKIMEDIA COMMONS.

pollination strategy so rapidly says a lot about how the relationship between bees and flowers actually works.

Any discussion of coevolution quickly runs into what philosophers call a "causality dilemma," a problem the rest of us recognize from the question, "Which came first, the chicken or the egg?" For bees and flowers, we know that both sides arrived at the party well-primed for dancing. Branched hairs apparently complemented a bee's taste for pollen from the earliest stage of their evolution, as Seán Brady pointed out to me: "All bees have them, so they must be as old as bees." On the botanical side, plants had long been experimenting with insect pollination, enticing potential suitors with nectar, or, more crudely, edible blossoms. (Some of these ancient strategies persist—Monet's famous water lilies, for example, would have flourished even if his garden hadn't had bees, since their pollinators also include small, flower-eating beetles.) Lack of fossil evidence makes it impossible to run the movie backward and watch the first steps of the dance unfold, but modern studies suggest that plants are often the ones taking the lead. When researchers tweaked monkeyflowers from pink to orange, for example, pollinator visits shifted from bumble-bees to hummingbirds in a single generation. A similar experiment on South American petunias showed that the flowers could trade in bees for hawk moths, or vice versa, by altering the activity of a single gene. These findings confirm that relatively simple steps in floral evolution can have dramatic consequences for pollinators, changing how some experts talk about the bee/flower relationship.

Turn to the relevant pages of a biology textbook, and you'll almost always find pollination described in glowing terms, with flowers providing nectar as a "reward" to their "beneficial" visitors. Scientists call this kind of reciprocity a "mutualism," the biological equivalent of win-win. But dig a little deeper and you'll notice some researchers using more clear-eyed language, words like "manipulation" and "exploitation." Because it's not all goodwill and handouts at the great floral buffet. Nectar, for example, is costly for plants to produce, and they don't just offer it up like a bowl of Halloween candy. Most flowers dole it out on a schedule, in locations and in amounts that direct precisely when bees visit, where they go, and how long they stay. And while that nectar does contain sugar, it's usually just sweet enough to be worthwhile, but not nearly as concentrated as what bees prefer. (After all, when they cook for themselves, bees make honey.) The range and ingenuity of pollinator manipulations is astounding. Some plants include caffeine in their nectar, inducing bees to remember and revisit flowers in a way that can only be described as habitual. Hiding nectar at the ends of spurs or tubes forces bees to plunge their heads deep inside to reach it, passing anthers and stamens along the way. Other flowers use pollen as the lure, or even edible oils, often tucking them into pores or pockets that force bees to linger in position, shaking and scraping their prize free. Pendant or upright flowers typically feature footholds or landing pads to position the bees, where even microtexture plays a role. Unlike the cells on any other plant parts, those at the surface of petals are cone-shaped, with the pointy ends sticking up. Eliminate these tiny structures in a lab, and bees start to slip and scrabble when they land, like dogs running on a hardwood floor. Floral traits that influence bees are as varied and widespread as bee pollination itself, but perhaps no group of plants pushes its visitors around more creatively than the orchids, whose flowers sometimes resort to outright trickery.

In the shady evergreen forests where I live, springtime arrives with a scattering of small pink orchids called calypsos. These thumbnails of coral brightness open just as the first queen bumblebees emerge from hibernation and begin to forage. The orchids emit an alluring fragrance and look like perfect bee fodder, with a broad landing area, beckoning stripes, and paired, shallow spurs that advertise the promise of nectar. Some varieties even boast anther-like hairs that glow yellow with the appearance of pollen. But the entire display is a ruse, and any bee drawn in gets nothing more for her trouble than two globs of pollen glued to her backside. The pollen is out of reach and stuck together in packets—useless to the bee, but perfectly positioned for delivery at the next calypso, should she fall for the ploy again. In fact, bees learn very quickly to avoid such false flowers, but when a single orchid can produce tens or even hundreds of thousands of tiny seeds, a little pollination goes a long way. Another spring bloomer, the lady slipper orchid, takes the deception strategy one step further. After luring bees in with scent, it traps them briefly in a deep, pouch-like lip. Disoriented, the bees are drawn toward clear "windows" at the back of the flower that lead to a narrow escape route. Pollen is deposited (or received) as the bees crawl up and out.

Fully a third of all orchids rely on some form of deception for pollination, and even those that do offer an honest reward often make bees go through gyrations to get it—dangling by their front legs, swimming through puddles, or tumbling down slide-like chutes. Male Euglossine bees in the American tropics do all of these things and more, visiting orchids not for nectar but to gather floral perfumes that play a role in their mating rituals. Hundreds of different orchids produce particular scents to attract particular species of bees, and their elaborate structures for pollen transfer are shaped and sized to fit those bees exactly, cementing the bond between flower and pollinator. A combination of scent and mating also defines what may be the most bizarre orchid strategy of all, a form of deception that was either too fantastical, or too improper, for early observers to even contemplate. During the great natural history craze of the nineteenth century, only one little-known amateur hinted at the truth.

Like his father and grandfather before him, the Reverend Ralph Price served as rector, patron, and vicar at Lyminge, County Kent,

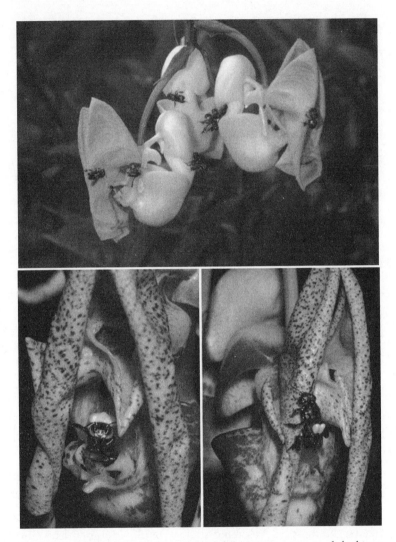

FIGURE 4.5. Male Euglossine bees in Central America swarm around the bizarre flowers of a bucket orchid in the genus *Coryanthes* (top). In gathering droplets of scent for their mating rituals, the bees slip and fall into the fluid-filled bucket, where they paddle around for as long as thirty minutes before discovering the escape hatch at the back of the blossom. Pollen is deposited (or received) as they drag themselves up and out to freedom (bottom left). Packets of pollen are often clearly visible on the bees afterward, while they rest and dry themselves off before taking flight (bottom right). PHOTOS © GÜNTER GERLACH.

in the south of England. His position offered a respectable living with enough leisure time for long rambles through the country-side in pursuit of his real passion, rare plants. As a botanist, Price distinguished himself by rediscovering an unusual member of the bellflower family, as well as by his observation that bees commonly "attacked" orchids in the genus *Ophrys*. Certainly, *Ophrys* flowers had long been subjects of interest, with their fanciful shapes that seemed to mimic the bodies, wings, and even antennae of insects. But no one had ever made such a claim before, and when Charles Darwin got wind of it, he was nonplussed: "What [Price's observation] means I cannot conjecture." Through some combination of scientific neglect and Victorian propriety, no one else ventured to conjecture on it either, for over half a century. But by the 1930s, multiple studies, from France to Algeria, had all come to the same conclusion: the bees weren't attacking the flowers, they were trying to have sex with them.

For an *Ophrys* orchid, successful cross-pollination requires three layers of cunning deceit. The first step involves attracting male bees (or in some cases wasps) with an odor that precisely mimics the scent of an available and amorous female. Next, their insect-like flowers cause the males to pounce and hold on tight, convinced they've found something with the size, shape, and smell of the real thing. To complete the ploy, dense hairs fringing the flower give the tactile impression of a fuzzy female, inducing the final act in what science has memorably titled *pseudocopulation*. By the time the un-suspecting suitor realizes his mistake, two pollen packets have been carefully placed on his head or abdomen, ready to be delivered to another flower the next time his libido gets the better of him.

For their part, bees don't exactly pollinate flowers out of gener-osity or some sense of botanical affection. They simply want nec-tar, pollen, or whatever other enticements are on offer, and will take them in the most efficient manner possible. Short-tongued

FIGURE 4.6. Bee orchids in the genus *Ophrys* mimic the scent and form of females, luring in male bees who unwittingly pollinate the flowers while attempting to mate with them. Clockwise from upper left: *O. bombyliflora, O. lunulata, O. insectifera, O. cretica.* PHOTOS BY ORCHI, ESCULAPIO, AND BERND HAYNOLD VIA WIKIMEDIA COMMONS.

bumblebees, for example, don't hesitate to chew through the base of a columbine spur or honeysuckle blossom, cutting a direct path to nectar that completely bypasses the flower's carefully crafted pollination scheme. (And once that hole is made, all sorts of other bees and insects quickly learn to use it.) Honeybees accomplish the same thing on mustards, not by chewing holes, but by sneaking up on the blossoms from behind and poking their tongues in through gaps between the petals. This kind of bee thievery is common enough that some botanists think it helped push the evolution of tightly clustered flowers with protected backs and bases, like those found in various clovers, mints, and members of the aster family. For bees that don't collect pollen, like the thousands of cuckoo species living off the labors of others, there is even less evolutionary incentive to play nice. They still take nectar from flowers, but many have done away with their pollen-gathering fuzz, becoming not only smooth and wasp-like in appearance, but waspy in their inefficiency as pollinators.

Even when bees do collect pollen and enter by the front door, they don't particularly go out of their way to be helpful—pollination is always less an intent than a consequence. Their interests lie in gathering and transporting pollen efficiently, an imperative that runs counter to spilling it carelessly all over the next flower they visit. Highly evolved groups like honeybees, orchid bees, and bumblebees carefully groom stray pollen from their bodies, moisten it with nectar, and pack it into dense, sticky clumps on their back legs. While this technique is great for moving pollen from blossom to hive, it renders that pollen useless to any flowers visited along the way. Such bees remain valuable as pollinators, but only inadvertently, because they miss a few stray grains while grooming, and can't see the stuff dusted across their backs. From an evolutionary standpoint, the bee/flower relationship is indeed special, but in strictly unsentimental terms: bees perceive flowers as a resource, and flowers use bees as convenient tools. The stark deceit and undiluted wants of pseudocopulation may say it best—all those male bees "attacking" *Ophrys* orchids excel at pollination without even realizing they've visited a flower.

History does not tell us whether Rev. Price suspected the significance of his bee observations, but experts now turn to the *Ophrys* example to understand how pollination strategies can lead to new species. The example is apt because, while questions about the parallel rise of bees and flowering plants date back to Darwin and Saporta, documenting that connection has proven surprisingly difficult. Any given bee/plant interaction usually takes place in the context of multiple pollinators, mimics, competitors, pests, and other factors playing out across a dynamic landscape, making it nearly impossible to tease out the effects of coevolution from the routine hum of individual adaptation. The time frame is also a challenge. With new lineages branching off on three to five occasions every hundred thousand years or so, *Ophrys* orchids have diversified as fast as any plants studied to date. But since the average graduate student spends only two to four years on a project, and a whole career flashes past in mere decades, speciation can't exactly be studied in real time. Most efforts have instead remained theoretical, relying on broad evolutionary trends, simulations, and the reams of circumstantial evidence offered by pollination syndromes. Only recently have genetic and conventional approaches combined to show just how new species can arise from pollinator interactions, with the specialized flowers of *Ophrys* as a prime case study.

It is at this point that the scientific literature on bees and plants begins to sound like a zombie apocalypse novel, peppered with words like "mutation" and "radiation." But unlike horror and science-fiction writers, biologists use these terms in a positive light. Mutations are simply random, heritable changes in genetic code that sometimes affect distinct traits, like the fragrance of a flower. Mutations provide much of the variation necessary for evolution, and favorable ones can sometimes trigger a rapid proliferation of new forms from a common ancestor. That process is called a *radiation*, in the sense of radiance, as if the new species were beams spreading out from a glorious source of light. Genetic studies of *Ophrys* show that small mutations can quickly alter scent production, luring entirely different kinds of male bees. The new bees immediately provide that vital ingredient for making a new species—reproductive isolation. Since they aren't attracted to flowers that retain the old smell, they collect and deliver pollen only among the orchids with the new one, immediately putting those plants on a separate evolutionary pathway. The exclusive nature of that bond (there are no other pollinators) makes for an unusually clear-cut evolutionary story: new scent begets new bee begets new species, with radiations occurring whenever the orchids chance upon a diverse new group of bees to exploit.

The *Ophrys* example highlights one of the main ways that pollinator relationships can lead to new species: specialization. Whenever an interaction becomes so idiosyncratic that the plants or bees involved stop mixing with others of their kind, new species can arise. *Ophrys* shows one side of that equation, how bees can affect plant diversity. Plants, in turn, can prompt new bees, but to do so they need to change not only how those bees forage, but how they breed. Female mining bees in the genus *Andrena*, for example, often become so dedicated to a particular type of flower that it's the only place their male counterparts can reliably find them. The blossoms act like highly selective pick-up bars, providing just the kind of isolation needed for speciation. Not surprisingly, perhaps, *Andrena* ranks among the most diverse of all bee genera, with over 1,300 species that often look nearly the same, having drifted apart from one another based solely on their fondness for different flowers.

Specialization lies at the heart of many patterns in bee and plant evolution, from the extravagance of orchid pollination to the intricate dance between tongue length and the depth of flower spurs. But it bears mentioning that most bees visit a range of flowers, and that most flowers attract a range of pollinators. Specialists may have the advantage of a dedicated partner, but they suffer the corresponding risks of dependence—lose one side of the relationship to disease, disturbance, or bad weather, and the other side may peter out as

PLATE 33.

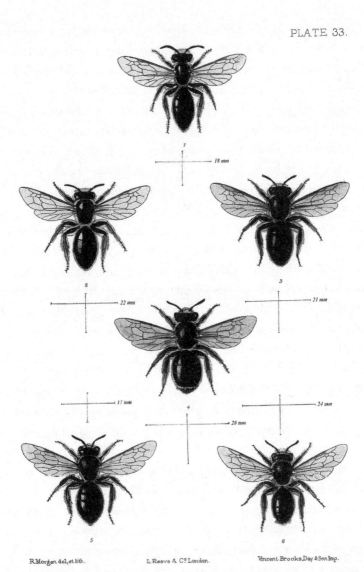

R.Morgan del, et lith. L.Reeve & Co London. Vincent Brooks,Day &Son Imp.

FIGURE 4.7. Six similar-looking mining bees in the genus *Andrena*, a group where new species sometimes branch off based only on their dedication to particular types of flowers. Pictured: *A. chrysosceles, A. tarsata, A. humilis, A. labialis, A. nana,* and *A. dorsata*. Edward Saunders, *The Hymenoptera Aculeata of the British Islands* (1896).

well. Being a generalist is like taking out a good insurance policy, and it's the dominant lifestyle for many diverse and highly successful plant families, like asters and roses, as well as for many bees, particularly the social groups with a lot of mouths to feed, including honeybees, bumblebees, and stingless bees. Evolutionary tension between the two strategies has also contributed to diversity. Because both approaches can be successful, the descendants of generalists often evolve toward specialization and vice versa, meaning that closely related bees or plants can vary tremendously in their foraging and pollination strategies.

The special relationship between bees and flowering plants doesn't explain all the species diversity in either group. For every case of pollinator-driven speciation, there is another instance where new forms arose through geography, range expansion, or rapid adjustment to new niches and environmental conditions. But no one doubts that pollinator interactions are fertile ground for the study of evolution. In fact, Charles Darwin followed up *On the Origin of Species* with a lesser-known volume titled *The Various Contrivances by Which Orchids Are Pollinated by Insects*. Sales were modest, but it's telling that he turned so quickly to bees and plants to find compelling illustrations of natural selection. And while *Origin* drew heavily on his far-flung travels aboard the HMS *Beagle*, many of Darwin's pollinator observations took place in his own garden and in the fields and woodlands nearby. That proximity is an apt reminder that while the coevolution of bees and flowers plays out over huge spans of time, its results and implications can be seen all around us. I've looked for bees everywhere from deserts to tropical rainforests, mountain meadows, and African savannas, but two of the most dramatic bee communities I know of occur within a day's journey of my island home. Together they helped me realize just what's possible when a landscape provides flowers and bees with everything they need.

CHAPTER FIVE

Where Flowers Bloom

Supply creates its own demand.

—Say's Law of Markets,
attributed to Jean-Baptiste Say (1803)

B umblebees rise early, and, as a toddler, so did my son. For the bees, getting a jump on the day gives them a chance to forage while most of their competitors are still abed, too chilly to fly. Bumblebees accomplish this feat by shivering heat into their flight muscles, an unusual ability we'll return to in Chapter Seven. As a warm-blooded mammal, young Noah's morning habits had nothing to do with body temperature. He simply regarded sleep as an inconvenience, something to be experienced grudgingly, in spans of a few hours at a time. Given this approach to life, there was nothing unusual about finding my little family awake and out for a walk in the early morning, surrounded by bumblebees.

We were visiting a small island not far from our own, where a good portion of my wife's clan lived in various cabins tucked into the woods. The trail was a familiar one, passing through a nature reserve

103

en route to the home of a sympathetic aunt and uncle who were also known to rise early, and who brewed strong coffee. Tall thickets of wild rose flanked the path, and my mind registered yellow-faced and black-tailed bumblebees lumbering among the pink blossoms. I vaguely wondered how many other species might be present, but mostly I was thinking about the coffee. It wasn't until we walked back that the bees stopped me dead in my tracks.

I've often told people that if you want to see more on a nature walk, don't take field guides, take a child. Noah's interest in bees had yet to blossom, but simply moving at his trudging, toddler's pace slowed me down enough to look closely at everything we passed. Warmed by morning sunlight, the rose bushes now pulsed with humming life. Bees seemed to festoon every flower, and they streamed through the air in a steady flow, darting around us as they zoomed by, as if the trail had been cut for their use alone. Holding Noah's hand and watching all the activity, I realized two things in rapid succession: (1) I had never seen so many bumblebees in my life; and (2) These were not bumblebees.

During the hours that we had been inside, visiting and drinking our coffee, the pollinator community along the path had been completely transformed. Sure, there were still a few bumblebees trying to elbow their way in among the roses, but the vast majority of that buzzing throng belonged to a species whose members only *looked* like bumblebees. I'd seen them just once before, on a collecting trip with experts from the US Department of Agriculture's "Bee Lab" in Logan, Utah. On that occasion, even the pros had been fooled at first. These mimics matched their bumblebee cousins almost exactly in size, shape, and yellow-orange fuzz—only their back legs gave them away. Where true bumblebees carried pollen in a basket-like structure on each tibia, the imposters stowed it in reddish fringes of brush-like hairs. That difference helped me recognize them as digger bees in the genus *Anthophora*, but what could explain their incredible numbers? This was normally an occasional bee, at best, but here they abounded in all directions, crowding a thicket that stretched from the shores of a nearby pond past the trail and right up to the edge of a high bluff that overlooked the bay. And that's when it hit me. Noah tottered ahead with his mother as I stopped, staring at the ground beneath my feet. Suddenly, I knew exactly where all those bees were coming from.

The Oxford English Dictionary traces the origin of the expression "Duh!" to a Merry Melodies cartoon from 1943. The similar term "Doh!"—popularized by Homer Simpson—got its start on a BBC radio program a few years later. Either phrase would have been appropriate for me in that forehead-slapping moment. As their name implies, digger bees dig, excavating nests in patches of bare soil, clay embankments, the walls of gullies and dry washes, or, when they can find them, the sheer faces of sandy cliffs. French entomologist Jean-Henri Fabre aptly dubbed them "those children of the precipitous earthy banks." For years I had been hiking that trail and idly watching bees on its prolific flowers without ever connecting the fact that it followed the crest of just such an earthy bank—a steep slope of sand and gravelly soil that rose to heights of fifty feet (fifteen meters) above the beach below. That afternoon, as soon as Noah was settled in for the restless period that counted as his nap, I grabbed a notebook and scurried down the beach for the first of many visits to a site still known in our family as "Papa's Bee Cliff."

When people visit the seaside they almost invariably gaze out at the ocean, drawn to the view by a measurable feeling of water-related calm that neuroscientists call "blue mind." That may explain why, in all my previous walks along that shoreline, I'd never noticed the half mile (eight hundred meters) of ideal bee habitat that lay just a few feet in the landward direction. (Perhaps a better name for my brain at the beach would be "blue mindless.") Approached now from below, the cliff rose like a wall of pale adobe, pockmarked here and there, but otherwise undistinguished. As with so many things about bees, little could be learned from a distance. Only after I clambered over the beach logs and stood right at the cliff's base could I see, hear, and feel the overpowering thrum of all those concentrated lives. If the path above had streamed with bees, then here they raged in a torrent, often crashing full speed against me in their haste to reach their nest holes. Scrambling up the toe of the slope, I found a place where I could sit and lean back against the warm sand while the frenetic bustle of an oversized bee community played out around me.

The first detailed description of this digger species dates to an article from 1920 by Harvey H. Nininger, who later gained fame for assembling the largest private collection of meteorites in the world. Apparently, the observational skills that helped him spot all those space rocks made him a good entomologist, too, and his account remains spot-on: "It was a bright spring day and the warm sunshine kindled the vital spark in these insects to the greatest activity. . . . They were engaged in digging tunnels, excavating nest-chambers, depositing eggs, and provisioning nest cells. All of these activities were being pursued most industriously."

I saw the very same behaviors and the same industrious vigor, but where Nininger estimated the population of his bee cliff in California's San Gabriel Mountains at around 100 individuals, I could see thousands at a glance. Up close, the pockmarks in the cliff resolved into a carpet of nest holes as dense as 60 per square foot (630 per square meter). But even so, the sheer number of bees outstripped the space available, and I saw tussles break out continuously as resident females struggled to fend off interlopers and keep control of their burrows. More than once, an entangled pair dropped right down on top of me and rolled off again, still fighting as they tumbled together down the slope. If these had been true bumblebees instead of look-alikes, I might have worried about getting stung. But while

FIGURE 5.1. A view of Papa's Bee Cliff, buzzing home to hundreds of thousands of digger bees, mining bees, leafcutter bees, and sweat bees, as well as all of their associated cuckoo bees, wasps, and other hangers-on. PHOTO © THOR HANSON.

diggers excavate their nests practically on top of one another, they remain fundamentally solitary, like mason bees, and lack the potent stings and coordinated defenses of social species. In fact, the digger variety at my cliff had taken their pacifist inclinations one step further. By mimicking a more dangerous species, they had engaged in a classic evolutionary bluff, adopting a threatening appearance that had become their primary means of defense. So long as bona fide bumblebees continued to sting, their digger mimics would be feared by association, allowing them to stop investing energy in their own defensive equipment and behavior. They still retained a stinging apparatus, but, as one observer noted, even with rough handling, "they cannot be induced to sting."

I lowered my head to the cliff face and watched a female as she reshaped the lip of her entrance hole, smoothing wet-looking soil with her abdomen until it formed a thin, raised lip. Like others nearby, the structure would eventually extend out an inch or two in length and curve downward in what Nininger described as a "peculiar bent-over chimney of clay." Some experts think the chimneys help hide nests from parasitic flies and wasps, while others suggest they might regulate nest temperature, or simply keep out raindrops and the loose dirt flung from neighboring excavations. Regardless of their purpose, they added a fascinating architectural element to the colony, which, from a bee's perspective, must appear like a vast desert city of wattle-and-daub turrets. That complex geography is a critical navigational tool for mama bees, helping them navigate back to their particular holes, legs dusted with pollen and honey crops brimming with nectar.

The life cycle of diggers resembles that of masons and other solitary bees, but instead of filling a straight tube with egg cells, they build a network of individual rooms that branch from the ends of their tunnels. The mama bees line each chamber with a cellophane-thin secretion that is both waterproof and resistant to

FIGURE 5.2. Digger bees invest considerable time and energy building elaborate, curved chimneys or turrets at the entrance to their nest holes, which may help keep them hidden from parasites, or protect them from weather. At the end of the season, some of the turret material will be repurposed as a plug to seal the tunnel. PHOTO © THOR HANSON.

rot, protection for the single egg laid within on top of a wet mixture of pollen and nectar. (Looking more like a slurry than the typical bee bread, digger provisions are sometimes referred to by the charming name "bee pudding.") All of this digging and provisioning meant that in spite of its apparent liveliness at the surface, the real activity of that bee cliff took place out of sight underground, in an unfathomable labyrinth of passageways and burrows. I couldn't peel back the soil to see what all those bees were up to, but at the very least I wanted to know how many there were. I'd never seen anything like it, which, in the world of field biology, often means you've stumbled onto something important.

My sister-in-law earned her doctoral degree in a laboratory, experimenting on bacteria, and she often teases me about the practice of biology out in nature. "All you people do is count stuff," she says. Like any good taunt, her comment includes a certain amount of truth. Over the course of my career, I've counted everything from seeds and fern spores to palm trees, bears, butterflies, gorilla droppings, and the pecking motions made by vultures. I now made a mental note not to tell my sister-in-law about the bee cliff project. She would never let me live it down if she knew I'd been reduced to counting holes. But while the methods were admittedly tedious, only painstaking measurements could override the visual confusion of all those swarming bees and provide an accurate estimate. Hole-counting at Papa's Bee Cliff became a regular scientific addendum to our family trips, and I could eventually say with confidence that at least 125,000 female diggers called the place home. The male bees lived nearby, staking out territories in the rose thickets and other patches of flowers where they waited for opportunities to mate. They generally outnumbered the females by at least two to one, putting the whole adult population at close to 400,000 bees on any given spring day. It was an impressive number, two orders of magnitude larger than other known populations of the species. But the more time I spent there, the more I realized that digger bees were just the beginning.

On that first afternoon visit I collected two specimens in an empty jam jar that I'd found on the beach, but in the future I never returned to the cliff without my favorite insect net, a collapsible model that could be popped out for a quick swipe pretty much any-where. Ever since my first netting lesson with Jerry Rozen, I'd come to realize that stalking bees was a vital aspect of understanding them. Just like walking with a toddler, pursuing something slowly and carefully sharpened the senses and created a whole new per-spective. At the bee cliff, I quickly noticed how the diggers clus-tered in soil of a certain grain size and density. Wherever things got too sandy or too hard-packed, different bees turned up—leafcutters, miners, long-horned bees, and sweat bees. There were sand wasps, too, and predatory tiger beetles patrolling the whole slope. Later in the season, cuckoo bees and a range of parasitic wasps appeared, sneaking in and out of the various nest holes whenever the mama bees were away. While diggers may have first drawn my attention to the cliff, its story turned out to be much more complicated—a whole community of insects exploited the nearby flowers, one an-other, and every available niche of nesting habitat. There were even things tunneling at the base of the cliff, where tailings from all the excavation above had piled up in a loose, slanting drift. I knew that teasing apart all those relationships required more than my general knowledge of entomology. To put names on all those bees, not to mention the wasps and flies and other species, I needed the help of an expert taxonomist. Luckily, I knew just whom to call.

I met John Ascher during The Bee Course, where he was the youngest instructor on the staff by nearly two decades. We bonded over bees, not surprisingly, but also music, after I overheard him improvising from memory on the battered old upright piano at the research station. He played beautifully, and when I mentioned that I was in a jazz band back home, he described his early struggle to choose between competing passions for music and entomology.

"After college I was hanging around in New York with a bunch of musician friends," he told me, and reminisced about the long jam sessions, and how they picked up gigs wherever they could find them. But as much as he loved jazz, he felt the others had something he lacked: "No matter how much I practiced, I could tell I'd never be as good as they were," he said, and then fixed me with an intense stare: "But I knew that if I concentrated on bees, I could be the best!"

By any measure, John was already well on his way. When we met, he had been working with Jerry Rozen for years, honing his skills as a curator of the vast bee collection at the American Museum of Natural History. He has since moved on to a professorship at the National University of Singapore, where he works on Asian bees and continues to identify North American species sent to him via Federal Express. (Happily, dried bees don't weigh much, and boxes labeled "Dead Insect Specimens" get waved through customs without a tariff.) John's brand of taxonomy is a fundamental aspect of the natural sciences—identifying species and learning how they relate to one another on the tree of life. But it doesn't receive much attention in an era increasingly dominated by technology-driven techniques and specialties. As more and more old-school taxonomists reach retirement age, the backlog of work keeps growing for young practitioners like John. Field projects often wait years to get their specimens professionally identified. But when I told him about the number of diggers at the bee cliff, John was eager to get involved. "I've seen that species in action," he wrote in an email, "but only dozens at a time."

In many ways, the bee cliff's extraordinary plentitude boiled down to a simple lesson in supply and demand. Biologist Bernd Heinrich explored similar ideas in his classic 1979 book, *Bumblebee Economics*. By tracing energy flows through the life cycle of a bee nest, Heinrich showed that inputs (nectar and pollen) bear directly on outputs (reproductive success). Increase the available flower resources, and a nest creates more bees. For diggers in the coastal environment where I live, suitable nesting cliffs usually occur in what amounts to a floral desert, with saltwater on one side and dense co-niferous woods on the other. But by simple good fortune, the acres of abandoned farm fields that topped my bee cliff had re-sprouted not with trees, but with a perfect suite of bee flowers—roses, black-berries, snowberries, cherries, and more. They bloomed sequentially throughout the spring and early summer, providing a profusion of nectar and pollen right alongside a vast expanse of nesting habi-tat. Energy in equals energy out, and the bee community had simply expanded to fit the available resources. In addition to the diggers, John identified ten other cliff and ground-nesting species from the specimens I sent him, as well as nine different cuckoo bees. All of these populations would be responsive to the same principles of floral economics. No wonder that path through the roses hummed with activity—it meandered through a diverse and prolific bee com-munity that must have numbered in the millions.

In nature, large bee colonies depend on the serendipitous oc-currence of abundant flowers and nest sites. Barring setbacks from disease or bad weather, those supplies will indeed create their own demand. The keepers of domestic honeybees have understood this relationship for thousands of years, gaming the system by moving their hives from place to place in pursuit of blooming flowers. This practice pays off not only in more bees, but in more of the golden honey they produce to feed themselves, and the wax combs they store it in, both of which can be harvested and sold. Just as impor-tantly, it allows for coordinated pollination on an industrial scale. When farms and orchards devote hundreds or thousands of acres to a single crop, it creates a brief and intense flowering period that often overwhelms local bee populations, particularly in highly cul-tivated landscapes with limited nesting habitat. The solution lies in a lucrative market for pollination services, and many commercial beekeepers now make more than half their annual income by rent-ing hives to farmers.

Throughout the spring and summer months, semitrailers stacked high with beehives crisscross the countryside, following a steady progression of bee-dependent crops, from almonds (which we will explore in Chapter Ten) to apples, pumpkins, cherries, watermelons, blueberries, and more. Like portable bee cliffs, the hive-laden trucks provide ample nesting habitat, while the succession of fields and orchards keep up a steady supply of nectar and pollen. The resulting honeybee populations can top 10 million individuals on the back of a single truck, a fact that few people appreciate more than the unlucky officers of various highway patrols called to the scene whenever one of these vehicles overturns. Traffic hazards aside, the long-distance conveyance of hives can pose significant risks to bee health, something we will discuss further in Chapter Nine. For some crops, at least, boosting native bee populations offers an attractive alternative. As Brian Griffin learned, mason bees nest readily in artificial blocks and pollinate fruit trees with abandon. Japanese apple growers now use them extensively. Certain leafcutters show similar promise, and there's growing evidence that simply maintaining undisturbed hedgerows can attract a range of bees and increase pollination for everything from blueberries to squashes. Even supposedly self-pollinating crops like soybeans seem to perform better with a mix of bees in attendance. Field trials continue, but one of the most successful native bee schemes of all time isn't a new idea at all. It dates back over half a century, to a small group of farmers in the American West, who I like to think fell for the charms of a particular bee just as hard as I had. As soon as I heard that alfalfa farmers had been building nesting beds for millions of alkali bees in the genus *Nomia*, I knew I had to go and see it for myself.

"You get more flowers, you get more bees. You get more flowers, you get more bees." Mark Wagoner gestured as he repeated this mantra, raising one hand and then the other like a pair of ever-climbing scales, as if visibly ratcheting up the size of his family's operation. After all, the principle had been holding true for generations. "My grandpa broke this place out of sagebrush," he told me as we toured one field, now lush with waist-high alfalfa in full bloom. Mark's son is a full-time partner in the business, too, and even his grandson seems to be off to a promising start—at two years old, the youngest Wagoner already ranks moving sprinklers among his favorite activities. That kind of long-term family commitment to farming is becoming rare in rural America, but it's not the only unusual thing about growing alfalfa in Washington's Touchet Valley, an irrigated oasis in the middle of the Columbia Basin.

"We use about 120 tons of salt," Mark explained, as we gazed out over another one of his fields. As a soil amendment, salt is generally reserved for sterilizing the cropland of one's enemies, but in this corner of his farm, Mark wasn't growing a typical crop. He was growing bees, and the salt formed a moisture-sealing crust over the soil, like it does naturally on the alkali flats that his bee bed was designed to mimic. Judging by the reaction of the bees, he seemed to have gotten pretty close to the real thing. They hovered and swarmed over the salted earth like heat shimmer, an uncountable frenzy of tiny bodies moving too fast for the eye to follow. It looked like my bee cliff, but tipped flat and multiplied tenfold. And instead of turrets, the bees had surrounded their nest holes with little conical piles of excavated dirt, like the tailings from thousands of tiny mines. But the biggest difference between cliff and bed had less to do with how the nests were arranged than with why. These bees were intentional, not happenstance, and Mark worked hard to give them whatever they needed.

"It's all sub-irrigated twenty inches deep," he said, pointing out rows of spigots and white PVC pipes that dripped in just the right amount of water, enough to keep the soil cool and firm for digging but not so much that it drowned the nests or made them rot. "The bees take priority," Mark added, and told me about a drought the previous season where the water district had cut off irrigation to crops, and people took short showers and let their lawns die. But the bee beds kept their allotment right through the peak of the nesting period. "They got water longer than anybody else," he said with satisfaction, sounding a little bit like a proud parent.

At that moment, my son, Noah, now a bee-crazy seven-year-old, successfully scooped up a buzzing female in one of the clear plastic vials we kept on hand for just that purpose. (In our family, this common catch-and-release activity had come to be known as "bee-tubing.") He held it up, and I immediately recognized the gorgeous opalescent stripes of my favorite bee. But it was hard to reconcile a species that I had glimpsed and collected only once, and had always considered rare, with the thrumming multitudes around us. Mark's bee beds, and those of his alfalfa-growing neighbors, embodied the cultural meme "If you build it, they will come." Covering a total of more than 300 acres (120 hectares), those scattered beds provided prime habitat to an estimated 18 million to 25 million nesting females, not to mention at least that many males looking for mates. With the exception of commercial honeybees, it all added up to the largest population of pollinators ever measured, a buzzing metropolis known among bee researchers as the Eighth Wonder of the World.

Our tour of the Wagoner farm soon explained how and why this one particular native species had become so critical to the business, but the first thing I learned was more fundamental: Mark Wagoner liked alkali bees even more than I did. "You can't have her. She's mine," he told Noah—seriously, but not unkindly—and we all watched the little bee fly out of Noah's tube and disappear instantly back into the humming throng. Later, inspecting the soil moisture at another bee bed, I overheard him curse himself when he accidentally dumped a shovel full of dirt onto a nest hole. For Mark, caring about alkali bees meant caring about each individual

FIGURE 5.3. Cars and trucks drive at a snail's pace outside the small town of Touchet in Washington State—not because of traffic, but to protect the native bees that their alfalfa crop relies upon. PHOTO © THOR HANSON.

bee. He'd practiced that ethic since he was Noah's age, when his father would send him down to the bee beds with a BB gun to drive away hungry birds. Since taking over the farm, he'd worked tire-lessly with neighbors and local leaders to make alkali bees a priority, not just for alfalfa farmers, but for the whole community. Road signs throughout the valley read "Alkali Bee Area" and set the speed limit at a strict 20 miles (32 kilometers) per hour. But Mark drove even slower, inching along as bees zipped past the windshield and warn-ing us, "Roll up your windows—you'll get bees in the car!"

At sixty-four, Mark had the kind of solid frame and tanned face earned by a life spent outdoors, and he wore his jeans, boots, and baseball cap with the ease of long familiarity. "We've got 1,200 acres [485 hectares] in alfalfa," he told me, nodding out at the dense rows of waist-high greenery. If he'd been growing the crop for hay, our story might end there, but alfalfa farmers in the Touchet Valley specialized in seed production, and that required pollination. Even from afar, Mark's fields glowed with little clusters of purple blossoms that filled the air with a pungent floral aroma. It must have been intoxicating for the bees, luring them from their nests to a bounty of pollen and nectar that stretched in all directions. But what happened when they reached those flowers wasn't just a simple extraction of rewards. Alfalfa blossoms hide their pollen and nectar inside folded petals that spring open when bees trip them, releasing the stamens and pistil with a surprising upward thrust. The result is a solid blow to the body or head that most bee species simply won't put up with. Honeybees, for example, quickly learn to bypass the punching parts of the flower by robbing nectar from gaps between the petals, leaving the blossoms untripped and unfertilized. But alkali bees don't seem to mind getting bopped, happily visiting flower after flower and apparently thriving on a diet of almost pure alfalfa. Once farmers in the Touchet Valley realized what these little bees were doing, they knew they'd found the perfect pollinator.

"I'd like to go back in time to the 1930s and look around for alkali bees," Mark said at one point, musing about an era just before alfalfa production took hold. "They had to be living around here someplace." A few natural bee beds still occur along the banks of the nearby Walla Walla River, the source of the valley's irrigation water, and some alkali bees do visit local wildflowers in the arid shrublands nearby. But the vast majority of the population appears to have shifted its timing to focus on alfalfa, which blooms later and longer than most plants native to the region. For bees, altering when they emerge from their nests counts as a major ecological adjustment, but Mark and other local growers have changed, too, adapting the way they farm to better suit their bees. They stay up late into the night to spray their fields after dark, when the bees are tucked safely into their nests. They constantly tweak the design and management of their bee beds, and collaborate with entomologists to study the results. They lobby state and federal agencies, and pool their resources to fund university research on bee-friendly pesticides. Mark's efforts recently earned him an award from the North American Pollinator Protection Campaign, something usually reserved for career academics, agency scientists, conservationists, or small organic operations. The Touchet Valley is now widely regarded as a case study for the use of native bees in an intensively farmed, high-production agricultural setting. But in spite of the attention and honors, Mark told me, he still feels like he's just scratching the surface with alkali bees. "There's a lot more that I don't know than what I do know."

At the end of our tour, Mark slowed his pickup and pointed out one of the open-sided sheds he calls his insurance policy. It also teemed with bees, but these were imported European leafcutters that Mark purchased annually as a hedge against bad weather, disease, a pesticide mishap, or other crises that might damage his bee beds. Cousins to the mason bees, leafcutters also nest in wooden blocks and paper tubes that can be shipped anywhere, and alfalfa farmers buy them by the millions, mostly from commercial growers in Canada. Like alkali bees, leafcutters don't seem to mind getting bopped by flowers, and in some places they are the main pollinators for alfalfa. To Mark, however, they're just not the same as the local species. "I buy them, but I don't love them," he said, and then tried to put his feelings for alkali bees into words. "It's different—they're like part of the family. . . . It's hard to explain." He paused for a moment, and then added, simply, "Alkali bees are the reason I'm an alfalfa farmer."

Driving out of the Touchet Valley, Noah and I stopped one last time to listen to the bees. With the car turned off and the windows down, they sounded like a great vibration, a low bowed note

droning ceaselessly over the fields. For Mark and other local farmers, that music was the sound of their livelihood and a backdrop to
their lives. It embodied not only the relationship between bees and
flowers, but another deep connection, to which we will turn in the
next section of this book—the vital and surprisingly ancient link
between bees and people.

Bees and People

But if thou wilt have the favour of thy bees that they sting thee not, thou must avoid such things that offend them: thou must not be unchaste or uncleanly, for impurity and sluttiness (themselves being most chaste and neat) they utterly abhor; thou must not come among them smelling of sweat, or having a stinking breath, caused either through eating of leeks, onions, garlic, and the like.... [T]hou must not be given to surfeiting or drunkenness; thou must not come puffing and blowing unto them, neither hastily stir among them, nor violently defend thyself when they seem to threaten thee, but softly moving thy hand before thy face, put them gently by.... In a word, thou must be chaste, cleanly, sweet, sober, quiet, and familiar; so will they love thee and know thee from all others.

—Charles Butler,
The Feminine Monarchie (1609)

CHAPTER SIX

Of Honeyguides and Hominins

No bees, no honey.

— Erasmus,
Adagia (c. 1500)

E very year, nearly two thousand members of the Society for Conservation Biology come together for a five-day conference, where they network, share their findings, and discuss the challenges of studying and protecting threatened species and landscapes. The meeting's location changes annually, but even an exotic venue can't alter the fundamental irony that it must take place indoors, in stuffy rooms, which is pretty much the last place that a bunch of field scientists want to be. After a day or two, antsiness sets in, and it's not unusual to spot groups of people piling into rental cars and skipping out to the nearest national park. Sometimes, however, the best things to see lie right outside the windows of the conference room.

When South Africa hosted the gathering a few years back, it took place at Nelson Mandela Metropolitan University, on the outskirts of Port Elizabeth. Aside from the main cluster of buildings, most of the school's 2,100-acre (830-hectare) campus remains in undisturbed *fynbos*, a dry, shrubby habitat named after the Afrikaans phrase for "fine bush." On the second afternoon, after I'd given my paper and answered a few questions, I was gazing out the window as the next session got underway. From a distance, fynbos looks nondescript, a slightly undulating expanse of sun-baked heath. But then I began noticing small patches of color scattered here and there among the greenery. The fynbos was in bloom, and I suddenly realized that I was in the right place at the right time to witness something spectacular. I excused myself immediately and darted outside. Anyone else looking out that window would have then seen me disappear into the bushes, searching for an interaction at the very root of the relationship between bees and people.

It didn't take long at all to find the bees. On the pinkish, phlox-like flowers of an unfamiliar shrub, I spotted droves of just the species I was looking for. Coming from North America, this sight alone was a rare treat: honeybees in their native habitat. At home, I couldn't help feeling conflicted about these fascinating creatures, my interest in their biology at odds with knowledge of their impacts on local species. By one estimation, the honeybees from a single domestic hive consume pollen and nectar that might otherwise provision 100,000 nest cells for diggers, masons, leafcutters, and other native bees. But here the honeybees were just where they belonged, flitting about in precisely the kind of arid, African environment that gave rise to their species—as well as our own. I watched while they nectared, sometimes two per flower, and then tried following a few when they took off, to see if I could trace them home to their hive. But it was no use—after a few steps I always lost track of their flight path in the dense shrubs. So instead I settled down to wait and listen, hoping for assistance.

If I were writing a novel, this is the moment where I would tell you that a brownish, robin-sized bird landed on a nearby twig,

FIGURE 6.1. The honeybee at home—a native *Apis mellifera* worker nectaring on a native ice plant flower in South Africa. PHOTO BY DEREK KEATS VIA WIKIMEDIA COMMONS.

chattering excitedly to get my attention. I would then describe how I followed that bird as it hopped and fluttered from branch to branch through the fynbos, leading me directly to the bees' buzzing home. That didn't happen, but the strange thing is that it could have. The Greater Honeyguide earned its name through exactly the behavior I described, ushering people to beehives with boisterous hopping, flapping, and an incessant cry that bird books describe as, "ke, ke, ke, ke, ke, ke, ke!!!" The bird ranges widely across sub-Saharan Africa, and wherever it is found, traditional honey hunters have learned to rely on its unique talents.

In one study, following honeyguides increased the rate of nest detection by 560 percent, and the birds consistently led hunters to nests that were larger and more productive than the ones they discovered on their own. After a beehive has been located and breached, the honeyguide benefits by feasting on leftovers and scraps—its specialized diet has resulted in an unusual ability to digest beeswax. As one early European observer noted, people customarily reward their avian helpers with a calculated gift of comb: "The bee-hunters never fail to leave a small portion for their conductor, but commonly take care not to leave so much as would satisfy its hunger. The bird's appetite being only whetted by this parsimony, it is obliged to commit a second treason, by discovering another bees-nest, in hopes of a better salary." Although no honeyguide materialized to help me in the fynbos that afternoon, its habits are a commonplace, well known to ornithologists and immortalized in one of the greatest scientific names of all time, *Indicator indicator*.

The first research paper on honeyguides was read to a meeting of the Royal Society of London in December 1776. It mentioned the bird's assumed natural counterpart, a mammalian hive-raider called the *ratel*, or honey badger. For over two centuries, common and scientific wisdom maintained that guiding behavior evolved between bird and badger, and that people had simply come along and learned to exploit it. It wasn't until the 1980s that a group of South African biologists pointed out what should have been obvious all along: honey badgers are almost entirely nocturnal. While their waking hours do overlap briefly with honeyguides at dusk, such limited opportunities hardly seem like a good coevolutionary starting point, particularly for such a complex interaction. Digging deeper, the doubters learned that honey badgers are nearsighted and hard of hearing, and that they rarely climb trees to the arboreal hives so often revealed by the birds. Playing recorded honeyguide calls to captive badgers produced no response, and it turned out that every published account linking the two species in the field was anecdotal, derived from hearsay or folklore. No biologist, naturalist, honey hunter, or even safari-going tourist had ever witnessed a bird leading a badger to honey. While the myth persists in natural history

FIGURE 6.2. People long assumed that the Greater Honeyguide (top) developed its remarkable guiding habits by leading honey badgers (bottom) to beehives, in spite of the fact that the bird is active by day and the badger is mostly nocturnal. Most experts now agree that the bird developed its remarkable traits in partnership with human ancestors. WIKIMEDIA COMMONS.

articles and even a best-selling children's book, finding the real story behind honeyguide behavior required biologists to go knocking on doors in a different department of science.

"My platform is nutrition," Alyssa Crittenden told me. "Everything builds from there. Diet is not where the story of human evolution ends, it's where it begins." I found Alyssa's door at the end of a narrow hallway in the anthropology building at the University of Nevada, Las Vegas. Her credentials include a prestigious named professorship as a nutritional anthropologist, but she has also studied ecology. That dual perspective helps Alyssa put questions about human eating habits into an environmental context. In conversation, she uses interesting phrases like "mapping people onto their food resources," and makes a convincing case that what our ancestors chose to eat helped define what we are today. If that's true, then people and honeyguides might have a great deal in common.

"If you want to study hunter-gatherers living in the same landscape where people evolved, that narrows things down pretty quickly," Alyssa told me, explaining how she'd begun her long association with Tanzania's Hadza tribe. Among the Hadza, roughly three hundred individuals live a strictly traditional lifestyle, ranging in small bands across the arid plains and woodlands that surround Lake Eyasi. Their homeland lies less than twenty-five miles (forty kilometers) from Olduvai Gorge and Laetoli, sites where fossils, footprints, and stone tools have documented the presence of human ancestors back more than three million years. Alyssa is quick to point out that groups like the Hadza are modern and culturally distinct. But as people getting their nutrition from a subsistence way of life in the same location where our species arose, they have a lot to teach us.

Alyssa spent her first season with the Hadza weighing and cataloging their daily harvest, from the fruits and tubers brought in by women and children to the various antelope, birds, and other animals hunted down by the men. She was interested in how seasonal fluctuations in food resources affected family life, particularly the women's decisions about when and with whom to have children. At the time, most nutritional studies in anthropology were preoccupied with what Alyssa calls "the meat versus potato debate," a long-standing clash about whether the calories from hunting or gathering contributed more to early human behavior and development. She suspected that there was more to the story, and like any good scientist, she kept her eyes and ears open. "I always follow the data," she said. But even Alyssa was surprised at the turn her research took when the data began pointing at honey.

"I was slack-jawed," Alyssa recalled, describing her first glimpse of a traditional Hadza honey hunt. She watched, fascinated, while the men scrambled up a massive baobab trunk on a series of rough wooden pegs, smoked out the hive with a torch, and brought down comb after comb dripping with golden honey. But that was nothing compared to people's reaction when the prize was brought back into camp. "The kids all started singing, and dancing, and horsing around. Everyone was so excited to share it, choosing out good bits to give to each other, and to me. It was unlike anything I'd seen before." The episode stuck with her and got her thinking. How much honey did the Hadza eat? Had she and her colleagues in anthropology failed to notice a significant source of calories? The more she looked into it, the more convinced she became. "Every foraging population for which we have data targets honey. Every ape species eats honey," she told me, retracing her thinking like a bulleted list. "It's nutritionally rich. It's highly preferred. Honey is an important food around the world, both now and in our evolutionary past. We had definitely missed something!"

Alyssa nearly missed it, too. Anthropology wasn't even on her radar screen when she started college. She wanted to be a doctor, and

FIGURE 6.3. A Hadza honey hunter poses with fresh comb from a hive of wild honeybees. PHOTO © ALYSSA CRITTENDEN.

was progressing nicely through the pre-med curriculum, when she happened to sit in on a course called Introduction to Human Evolution. "It made my mind explode," she told me, recalling how the class seemed to tie together everything she'd been thinking about. Intellectually, she described her abrupt change of career paths like Alice going down the rabbit hole to Wonderland. "I had too many burning questions," she said, and if our conversation was any indication, she still does. Surprisingly young for all her accomplishments,

Alyssa projects the sort of trim fitness and boundless energy that you might expect from an expert on nutrition. In two and a half hours, interrupted only by a trip to the campus coffee shop, we covered topics ranging from the chemistry of honey to the fletching on Hadza arrows to the challenges of academic editing. She often seemed just as curious about my work as I was about hers, asking questions with a combination of persistence and friendly informality that helped explain how she'd learned so much from the Hadza.

"Honey is their number-one ranked food," she told me. It was consistent in every interview she conducted—women and men of all ages, not to mention children, put honey far above any kind of fruit or meat as their absolute favorite thing to eat. Men and older boys searched for it daily, raiding not just honeybee hives but also the nests of at least six different stingless species. Women gathered some stingless honey as well, though by custom they didn't carry the axes necessary to breach the larger nests in trees and stumps. When Alyssa and her colleagues added up the data from years of observations, they found that honey provided fully 15 percent of the calories in the Hadza diet. "And that's an underestimate," she cautioned, since it didn't include all the nutrition found in bee larvae and pollen, which were also consumed with enthusiasm. Nor did their measurements take into account anything eaten outside of camp. For men, the calories from honey would have been much higher, since they typically gorged themselves whenever they found it, eating anywhere from a third to three times what they brought home to share. "They always complained about getting thirsty out there," Alyssa said with a laugh, noting how the body requires an influx of water to process all that sugar. "Just like my daughter on Halloween." But where trick-or-treaters only get to indulge their sweet tooth one night a year, the Hadza go looking for honey every day. And if our ancestors did the same thing when they occupied that habitat, it may explain a lot, starting with the bizarre habits of honeyguides.

FIGURE 6.4. Hadza honey hunters target the nests of seven different native bees, including members of the genus *Hypotrigona*, who build elaborate resinous tunnels at the entrances to their hives. A stingless species, their name in one regional dialect has been translated as "peaceful small insects that visit coffee flowers." PHOTO © MARTIN GRIMM.

"I'm actually not all that interested in the birds," Alyssa admitted, although the Hadza do follow them whenever they get the chance. But while she leaves it to others to study how honeyguides and humans interact, Alyssa's research leaves little doubt about how it all got started. She champions the idea that our taste for honey came from deep in our primate past, a notion supported by the fact that all living great ape species also seek it out. If honeyguides evolved three million years ago, as genetic evidence suggests, then they appeared on the scene in East Africa when our ancestors were already well-established residents of woodlands and savanna, leaving their bipedal footprints all over the neighborhood. In that context, why would proto-honeyguides have bothered trying to attract the attention of a nocturnal badger? The accepted theory now is that the bird coevolved with those early, upright hominins, who were already out there in plain view, looking for honey all day long. The fact that modern honeyguides focus their attentions exclusively on people is not surprising—it's a trick they've been practicing on the genus Homo for eons. But to Alyssa and other nutritional anthropologists, the most intriguing part of the honey story doesn't concern birds at all. It has to do with a critical evolutionary step that helped define our species.

"The brain is an obligate glucose consumer," Alyssa said, reminding me of a basic lesson in human biology. Because the brain burns energy for neurotransmission as well as basic cell function, its tissues are what physiologists call "metabolically expensive." Though an average human brain accounts for only 2 percent of bodyweight, it can eat up 20 percent of our daily energy requirements. And it demands all that power in the form of glucose. To keep up, the body breaks down starches from the foods we eat, or, with a little help from the liver and kidneys, reorganizes the energy found in proteins and lipids. But no natural food in the human diet contains more glucose, in a more unadulterated and digestible form, than honey. A full one-third of the calories in a spoonful of honey is pure glucose, with much of the balance coming in the form of fructose, a similar sugar. "It's the most energy-rich food in nature," Alyssa observed, and the need to feed our big, hungry brains may help explain why we crave it.

Any good textbook on human evolution features an image of the skull known as "Nutcracker Man," a specimen from the genus *Australopithecus* discovered near Olduvai Gorge by Mary Leakey in 1959. It looks nearly human, though with a relatively small brain-case and a jutting lower jaw filled with massive molars—the inspiration for its nickname. In contrast, skulls from the genus *Homo* are distinctive even to the untrained eye by their combination of smaller jaws and teeth, a flatter face, and a lot more room for gray matter. Sudden leaps in brain size are the hallmark of our pedigree, to the point that a modern human boasts two and a half times the brain capacity of old Nutcracker. For nutritional anthropologists like Alyssa, every change in those ancestral skulls raises important questions about diet. Early humans could never have afforded the metabolic expense of larger brains without an accompanying boost in calories. The shift to smaller dentition tells part of the tale, suggesting a transition to softer, richer foods. Most theories to date give the credit to increased meat consumption through hunting, or to the advent of tools for acquiring and preparing tubers and other new foods. Control of fire introduced the nutritional advantages of cooking, another potential factor. To this list of dietary innovations, Alyssa and her colleagues have added honey, the most potent brain food of all.

"There's momentum now," Alyssa told me at one point. "Honey is gaining traction." She explained how until very recently, ancient honey consumption had been impossible to document. Unlike other dietary habits and advances, it didn't leave behind distinctive tools, charred hearthstones, or the telltale marks of butchery on bone.

This may be another example of preservation bias, overemphasizing events that happened to leave behind a clear trail of artifacts. Until recently, honey was overlooked because it simply couldn't be seen. But new techniques can now pinpoint lingering chemical finger-prints from even the tiniest stains and residues. Studies have already turned up convincing evidence of beeswax from thousands of pot-sherds, as well as what appears to be the world's first dental filling, confirming a strong honey connection at the dawn of the Neolithic. For the more ancient time period that Alyssa is interested in, she rests her hopes on something that anthropologists used to think of as a blemish—dental plaque.

"We always used to wash tooth specimens," she said, making a scrubbing motion with her hands, "but now we know better." While a dirt-free fossil might look good in a museum display, cleaning it also removes critical data lodged in its nooks and crannies. Fossil plaque contains a surprising amount of information about ancient diets, and it can even hint at social behaviors. The recent discovery of distinctly human oral microbes in the plaque of Neanderthals, for example, raises the possibility that the two species once shared meals together, or, more controversially, locked lips in a juicy pre-historic smooch. Alyssa is convinced that analyzing plaque from the right time periods will turn up honey traces from all the key points in our evolutionary history. Like hunting animals, finding honey provided our ancestors with a rich nutritional reward for complet-ing a complex task. It would have created a similar impetus for the development of cooperation and sharing, as well as tool use and the mastery of fire. The hand axes, flakes, and other stone implements that led to efficiencies in killing and butchering game, would also have allowed access to the larger bee nests hidden in trees. And while fire may have given us a nutritional boost through cooking, so too would it have allowed the pacification of honeybees with smoke. If our ancestors did indeed search for honey as regularly as the Hadza

FIGURE 6.5. A wild honeycomb collected by the Hadza is a nutritional windfall, supplying a boost of sweet calories from the liquid honey, as well as protein and nutrients from cells filled with bee larvae and pollen. PHOTO © ALYSSA CRITTENDEN.

do today, then each of these advances would have been accompanied by a huge surge in sugary calories. And, as Alyssa reminded me several times during our talk, bee nests also contain larvae and pollen, which provide additional calories as well as protein and important micronutrients. Taken together, these dietary contributions make a strong case that learning to follow bees (and honeyguides) influenced human evolution, helping our ancestors to bolster their growing brains, and—in the language of anthropology—"nutritionally outcompete other species."

People will always debate the factors that made *Homo sapiens* so large-brained and dominant, but Alyssa and her colleagues have succeeded in giving honey a seat at the table. Their theory quickly found a place because it complemented, rather than replaced, the existing paradigm. Nobody believes that eating honey is what made us human, but few experts now doubt that it was a valuable and nutritionally potent part of our ancient diet. While the idea initially attracted me because of what it said about our connection to bees, I also came to admire how Alyssa and her colleagues developed it—moving from an intriguing observation to a simple proposal to something more sweeping. It's all there in the list of publications available on Alyssa's website, outlined in the way her coauthors and topics have proliferated over the years, from honey and digestion to the wear patterns on stone tools and tooth enamel. (Poring over other people's bibliographies is one of the nerdier pleasures of a career in science.) Alyssa wrapped up our conversation by repeating something she'd said at the beginning—the fundamental question that ties together all of her work: "How did we end up walking around in bodies like this, living like we do?" Then she was off to pick up her daughter from preschool, which reminded me of another trend in her research: a series of papers on the foraging habits of Hadza children.

It should come as no surprise that young hunter-gatherers have a sweet tooth. Children everywhere exhibit a measurably higher tolerance for sugar than adults, particularly during periods of active bone growth, when their bodies yearn for the quick energy from easily digested calories. Young Hadza begin by targeting figs, berries, tubers, and baobab fruits close to camp, but they soon learn that several varieties of stingless bees build their nests within easy reach—low down in hollow tree branches, or even underground. When boys get old enough to wield an axe, a traditionally masculine tool, they graduate to tree-nesting bees, and eventually begin following honeyguides to the largest and richest hives of all. Much of that sweet treasure gets consumed right on the spot, possibly helping to fuel the rapid growth spurt common in adolescent males. That combination—sugar cravings and growing bodies—may help

FIGURE 6.6. Sugar cravings peak measurably in children during periods of active growth, when the body craves easy calories. Before the advent of inexpensive refined sugars, like the dextrose promoted in these vintage advertisements, kids in rural areas everywhere commonly fed their collective sweet tooth by seeking out the nests of wild bees. IMAGES COURTESY OF SALLY EDELSTEIN COLLECTION.

explain why children around the world continued searching for the nests of wild bees long after the practice disappeared from their cultures at large.

Domestication brought honeybees into agriculture early on, largely eliminating the need for regular honey hunting. But while their hives could be readily maintained on a farm, honeybees remained defensive and fierce, requiring smoke and other techniques that left their management (and their honey) largely under the control of adults. The more gentle species, however, remained vulnerable to anyone looking for a quick, sweet treat, and, until quite recently, children in rural areas everywhere knew their habits well. Famed French entomologist Jean-Henri Fabre traced his fascination with insects not to a textbook or a university class, but to an experience watching schoolboys rob the sweet provisions from mason bee nests. In Japan, children compared the taste of bee bread to a popular confection made from soy flour mixed with honey. One common mason species is still known locally as *Mame-ko bachi*, "the bean-flour bee." Bumblebees made an even more attractive target, their modest but tasty cache of liquid honey worth the risk of a few stings. During the nineteenth century, raiding bumblebee nests was such a standard childhood activity that it even appeared in poetry, such as this rhyme from the popular collection *Jingles and Joys for Wee Girls and Boys*:

> *Rig a jig, you bumble-bee,*
> *Make some honey sweet for me,*
> *Then fly away to make some more,*
> *To add to that you have in store . . .*
> *And just as sure as you're alive,*
> *I'll make a visit to your hive,*
> *And thank you bumble-bee so bold,*
> *For making honey bright as gold.*

The practice remained common until at least 1909, when the following anecdote appeared in an article promoting bee observation as a classroom science project: "Yesterday morning a boy came into my office to tell me of the big bumble-bees' nest the boys had just robbed, and of the quantity of bumble-bee honey they had taken. . . . [I]f there is one insect in this country that the ordinary country or small town boy knows more about than another, especially at the season when the second crop of red clover is in bloom, it is the bumblebee."

By the latter part of the twentieth century, however, something had changed. My own upbringing as an "ordinary country or small town boy" in the 1970s did not include a single experience with native bees. Never did I pluck the bread from a mason nest or join my friends on a bumblebee honey raid. When we wanted something sweet, we did what all the other kids did—we bought candy. Through some combination of changing attitudes and the ubiquity of refined sugar, children of my generation—even those interested in nature—had lost the urge to seek out bees. Now, as a middle-aged biologist, I suddenly found myself wanting to make up for lost time. And when my own son reached an age when Hadza children would start learning to forage, I discovered that I had an accomplice.

CHAPTER SEVEN

Keeping Dumbledores

There are certain pursuits which, if not wholly poetic and true, do at least suggest a nobler and finer relation to nature than we know. The keeping of bees, for instance ... is like directing the sunbeams.

—Henry David Thoreau,
"Paradise (to be) Regained" (1843)

"I hear a bee!" Noah shouted, looking up from his excavator. Like many young boys, Noah felt a strong devotion to toy trucks and earthmovers, and he'd spent the past hour patiently leveling a patch of muddy ground in front of my office. (I work in a converted shed in our orchard that we call "the Raccoon Shack" in honor of its former occupants.) It pleased me immensely that a bee could divert his attention, but, after all, we'd been waiting to see one for days.

We both froze as the bee rounded a corner of the shack and began to inspect the porch. It started with a knothole in the siding, and then moved up to the eaves, bumping along a narrow shelf I'd built as a nesting site for swallows. I found myself holding my breath as the bee worked its way downward again, closing in on an odd contraption nailed to the porch's lattice screen. Just as a good dry ledge might attract a swallow, Noah and I hoped that our peculiar wooden box would prove irresistible to bees. Past failures had led to this season's innovation: an old rain boot tacked on as an entry tunnel, its cut-off toe stuck tight to a hole in the side of the box, and its gaping top beckoning outward toward the orchard trees. The bee hovered for a moment in midair, halfway between the eaves and the lattice. Then, as if yanked by a strange gravity, it lurched forward and flew straight into the boot.

"Was it a *Bombus*?" Noah asked eagerly, using a Latin name often overheard in our increasingly bee-crazed household. I nodded yes. Identifying bees is usually more challenging, involving pinned specimens, a dissecting microscope, and a clear view of features like wing veins, tongue length, or, in some cases, the pattern of notches and grooves on the male genitalia. But when you spot a bee on the wing there is one good rule of thumb: if you're wearing a wool hat, two layers of flannel, and a down vest, you're looking at a bumblebee. Few insects are so well adapted to cool weather, with the ability to temporarily unhook their wings from their flight muscles and shiver the muscles alone, generating heat in the thorax and pumping it throughout their fuzzy, well-insulated bodies. This skill allows them to reach flight temperature in a wide range of conditions, and I knew that nothing else would be flying on such a blustery afternoon. And since it was only the second day of March, I knew that the bumblebee had to be a queen, newly awakened from hibernation, braving the cold to search for a place to start her colony.

A muted buzzing marked the bee's progress as she made her way down the boot, through the toe, and into the box. I tried to picture her there in the darkness, encountering our various enticements by scent and feel: the cotton batting for lining her nest, the thimble-sized cup of fireweed honey. British entomologist Frederick

William Lambert Sladen used to equip his boxes with everything from hand-cut grass to shredded flax fiber, and went so far as to feed bees with an ink dropper. He even fashioned artificial honeypots for them, shaping molten wax over "the rounded end of a wooden stick previously moistened with water." Such detailed suggestions make his 1912 book, *The Humble-Bee: Its Life History and How to Domesticate It*, a primary resource for any aspiring *Bombus* keeper. In the century since its publication, people have largely abandoned the charming name "humble-bee," and the even older variation, "dumbledore," is now known only to fans of Harry Potter. But we still keep bumblebees close to us. Entomologists call them the "teddy bears" of the bee world, and, like honeybees, some species have become important commercial crop pollinators. They are exceptionally good at what experts call *sonication*, or *buzz pollination*, vibrating their wings at just the right frequency to shake pollen loose from difficult flowers like those found on tomato plants (something we'll discuss further in Chapter Nine). But if Sladen were still alive, his first question for Noah and me would probably not have been about developments in bumblebee science. It would have been about the boot.

In nature, bumblebee queens seek out abandoned mouse or rabbit burrows, rocky clefts, hollow logs, or the tree cavities left behind by woodpeckers. They need a dry, enclosed place with enough room for a colony that might grow to several hundred individuals by the end of the season. Finding a suitable location requires incessant searching for the one feature that all such locations have in common: a dark hole for an entrance. This imperative gives bumblebee queens an insatiable curiosity about shady gaps and openings, places that have always abounded in the human world. An old Wiltshire saying likens the sound of mumbled speech to the buzzing of bumblebees inside a pitcher, which implies, of course, that finding bumblebees in pitchers was once a common experience that anyone could relate to. In fact, people have discovered bumblebee nests in all sorts of

FIGURE 7.1. With their long, dark openings and cozy toes, boots make an excellent place for bumblebee queens to nest, particularly when you leave them carelessly overturned on your porch in the springtime. Image from Paul Augé, *Larousse du XX siècle* (1928).

unexpected places, from teapots and watering cans to downspouts, chimneys, tailpipes, and rolled-up carpets. To this list I added rubber boots in the obvious way: by shoving my foot into one and receiving a ferocious sting.

The incident occurred right on the Raccoon Shack porch, where I'd left my muddy footwear unattended for several hours while working inside. (Knee-high rubber boots complement my office attire for much of the winter and spring, when the path from house to shack lies partially submerged in muck.) Chancing upon this dark and comfy location, the queen had apparently liked it well enough to start setting up housekeeping. That is, she liked it until my offending toe came barging in and spoiled everything. I kicked off the boot and saw her tumble out and fly off in search of more hospitable quarters, but in spite of my pain and surprise, the sting gave me a sudden feeling of hope. Perhaps I had finally learned how to attract queens to a nesting box! For years, my every attempt to establish and observe a bumblebee colony on the Raccoon Shack porch had failed. It seemed like an ideal location—quiet, shady, and surrounded by flowering fruit trees and berry bushes. As an added bonus, the floral bonanza of my wife's sprawling garden was only a short flight away. But although I had tried everything from drain tile to flower pots to a cardboard box accessed via garden hose, no passing queen had, to my knowledge, even slowed down for a second glance. The previous season, Noah and I had resorted to capturing fresh queens and transferring them directly into a fancy observation box purchased from Brian Griffin's company, only to watch each and every one of them fly away at the earliest opportunity. But now, within two days of attaching a boot to the entry hole, we had lured in a potential tenant.

The buzzing grew suddenly louder and the queen emerged, flying in ever widening circles around the boot, the lattice screen, and the porch in general. "She's memorizing the location," I whispered to Noah. Bees use a range of visual cues to navigate, including polarized light and the position of the sun, but mounting evidence suggests that their tiny brains are also capable of memorizing detailed mental maps of their surroundings. Bumblebees and honeybees transported far from their nests in dark boxes have made their way home again from distances of more than six miles (ten kilometers), and an orchid bee once made such a journey from a distance of fourteen miles (twenty-three kilometers). Patient, circling flights allow bees to identify and remember key landmarks, helping them pinpoint the locations of a nest or a good source of food. Rearranging those landmarks can stymie a returning bee, at least temporarily, as Brian Griffin had shown me with his mason nests. I wondered briefly whether having bumblebee neighbors meant I shouldn't dare move the lawn rake, ladder, deck chair, or other things stored on the porch. Just then the queen darted off, crossing the orchard and disappearing out over the pasture, straight into the brisk wind. But she returned moments later, as if testing her mental map, and continued to inspect the booted box. I grinned and gave Noah a high five. We were off to a good start.

The roster of notable beekeepers begins with Aristotle and Pythagoras, picks up the likes of Augustus, Charlemagne, and George Washington along the way, and continues into modern times with a rash of celebrities, including Henry and Peter Fonda, Scarlett Johansson, and Martha Stewart. In the literary realm, Virgil kept bees, and so did Tolstoy, who spent two full pages of *War and Peace* comparing the evacuated city of Moscow that awaited Napoleon's army to a "dying, queenless hive." Sir Arthur Conan Doyle had no bees of his own, but he did imply that beekeeping was the only activity stimulating enough to occupy the intellect of Sherlock Holmes in retirement. Recalled for one last case in the story "His Final Bow," Holmes raves about his bees in an aside to Watson, saying: "I watched the little working gangs as once I watched the criminal world of London." The beekeeping canon ranges from Shakespearean metaphor to scientific memoir to practical handbooks, but nearly every historical and literary account refers to only one species: the honeybee. In choosing to keep bumblebees, Noah and I had set off on a path far less traveled, and far less celebrated. In fact, there is really only one truly famous person who ever wrote knowledgably about the genus *Bombus*, and few people even realize that she did so.

In the last year of her life, Sylvia Plath kept honeybees in the typical fashion and wrote several poems about them, but her earlier work is filled with bee metaphors and references of a different variety. She is surely the only major literary figure to use the word "hibernaculum" in a poem, correctly referring to the shallow burrow where a pregnant queen bumblebee spends the winter. Plath gained this familiarity in the most natural way, by spending her childhood in the company of North America's greatest bumblebee expert. While literary critics know Otto Plath as an ominous presence haunting his daughter's poetry, entomologists remember him fondly. His classic book, *Bumblebees and Their Ways*, stands as a sort of American counterpart to Sladen's work, and it's obvious that some of his knowledge rubbed off on young Sylvia. Childhood friends remembered her as a keen naturalist, and her writing includes entomological asides about everything from solitary bees to parasitic, gall-dwelling wasps. In the autobiographical story "Among the Bumblebees," she recalls the character based on her father cheerfully capturing stingless drones that buzzed harmlessly in his fist. I wasn't sure what Noah might remember about our own bee adventures together, but I knew the bumblebee chapter would be short if that queen didn't succeed in building up a colony. Unfortunately, things quickly took a turn for the worse.

The key feature of our nest box, aside from the boot, was a clear Plexiglas window built into the lid. Removing a wooden cover gave us a clear view of everything going on inside without disturbing the inhabitants (or letting them out). The first time we peeked in, we saw movement in the cotton batting. It was the queen, rearranging things to her liking. I quickly replaced the lid and told Noah that we shouldn't look again for a few days until she had settled in. With luck she would soon start making honeypots and laying eggs, and the box would begin to hum with the buzz of many wings. Soon after, however, I heard quite a different sound from the Raccoon Shack porch—the distinctive, scolding racket of a house wren. When I investigated, I found the bee gone and the boot stuffed full of twigs, the beginnings of a wren nest that would eventually fledge a full brood of six noisy chicks. I like birds, and tried to be philosophical about the setback. But Noah was irate, pledging eternal enmity to the entire wren tribe for evicting our precious queen. As if to add insult to injury, we learned later that the wrens had made another assault on the local bee population. At the end of the season, when we pulled their stick-and-feather construction from the boot, we found its delicate cup lined with a felt of linty fluff that could only have come from the picked-apart nest of a wool-carder bee!

While the wrens had foiled a first-rate bumblebee opportunity, Noah and I did at least learn something from the experience. By the time spring rolled around again, regular trips to our local thrift house had supplied us with old boots of every description, as well as teapots, water bottles, and sprinkling cans—enough potential apartment space, we hoped, to satisfy the wrens *and* entice another queen. In a way, this was really just a bumblebee variation on an age-old trick for attracting honeybees. When traditional hunters like the Hadza chop open hives in trees, they often repair the damaged trunks with stones or mud, hoping bees will reoccupy the same sites again and again. (Repaired hives offered the hunters two clear advantages: they knew exactly where to find them, and if more bees did move in, they were easy to crack open again.) Early African beekeepers simply took the next logical step: capturing wild honeybee swarms by setting out hollow logs in promising locations. That practice continues in many rural areas, maintaining some African honeybee populations in a curious state of perpetual semi-domestication.

Honeybees can swarm whenever a hive grows large enough to produce new queens and divide, sometimes more than once in a year. But attracting bumblebee colonies can only happen in the spring or early summer, when queens are first emerging and establishing their nests. That contrast in seasonality is deeply rooted and explains many of the differences between these two familiar kinds of bee. Honeybees evolved in tropical and subtropical climates where hives can persist year round, with huge populations that need sophisticated forms of sociality and communication to maintain order. Bumblebees, on the other hand, are almost entirely temperate, adapted to living in places with harsher winters where the best survival strategy lies in the hibernation of queens. Their biology emphasizes immediacy, with workers switching among various tasks

FIGURE 7.2. A traditional form of African beekeeping involves attracting wild swarms to hollow logs or other inviting homes. In this picture from Ethiopia, dozens of potential hives dangle like birds' nests from an acacia tree. PHOTO BY BERNARD GAGNON VIA WIKIMEDIA COMMONS.

and social roles as needed to maintain productivity through what can be a very brief season. Alpine bumblebees, or those nesting in the arctic, must complete their entire colony cycle in a matter of weeks. This inherent brevity also explains why Noah and I had relatively little company in the world of bumblebee domestication. Because honeybees evolved to live year round, they produce massive quantities of honey, enough to maintain tens of thousands of workers through dry seasons, cold snaps, monsoons, or any period of time when flowers might be scarce. Bumblebees make honey, too, and it is equally delicious. But they produce a comparative pittance, just enough to feed a few scores of bees on the occasional rainy day.

The weather on our island improved as spring progressed, and Noah and I had high hopes for the boots and teapots scattered around our orchard. But as a back-up measure I also began reading everything I could find on the art of bee tracking. In places that lacked the convenience of honeyguides, hunter-gatherers had learned to catch foraging workers at flowers, and glue petals, leaves, or even feathers to their backs, making them easier to spot and follow on their flight path back to the nest. Careful listening was also an important tracking tool. Groups of Mbuti honey hunters in eastern "Congo" reportedly used hearing alone to pinpoint the buzz from between two and three nests *per person* on every outing, all within a short walk of their encampment. Such statistics gave me hope—if the bees chose not to set up housekeeping in one of our nests, then surely, over the course of an entire season, we could figure out a way to find one of theirs. As it turned out, that was easier said than done.

Two days of early spring sunshine brought us the much-anticipated first bee sightings of the year, but then the rains returned, battering our island with a series of cold, windy storms. Coming on the heels of a particularly damp winter, the weather seemed downright spiteful, even to those of us born and raised in the Pacific Northwest. But for the bees it was far worse. Awakened from their torpor, those early queens found themselves burning through precious energy reserves in a chilly landscape with hardly a flower in sight. Day after day we found bedraggled, wet bumblebees clinging to the occasional crocus or daffodil that had dared to bloom. Things warmed up again eventually, but that false start had taken its toll. In all of our nests we found only one early bumblebee, a black-and-orange-rumped queen who had crawled into a boot on the Raccoon Shack porch and died near the toe, overcome like so many others by some combination of cold, wet, and starvation.

Luckily, not all bumblebees are created equal. As Sylva Plath knew very well, a queen emerging from her hibernaculum marks more than just a beginning—she is also a continuation. The number, condition, and health of springtime bumblebees all depend directly on the successes and setbacks of the previous summer. At the end of every season, the old queen, the workers, and the drones all die, pinning their collective hopes on a few chosen survivors. In Bernd Heinrich's economic terms, those overwintering young queens represent net profit, the reproductive proceeds from all the effort and floral energy invested by their nestmates. New queens—and the drones they mate with—are produced late in the season, in whatever numbers their colonies can afford. A nest with few resources, or one suffering from parasites or disease, might not yield a single mated queen. But in times of plenty, colonies grow large enough to deliver them by the hundreds. Prosperous bees can also provide their queens with more food, producing large, robust individuals better able to survive a harsh winter or an unseasonably cool spring. Finally, bumblebees have evolved to hedge their bets. Variation in the cues that bring them out of dormancy prevents any particular generation of queens from waking up all at once—insurance against bad weather, erratic blooms, or other potential problems. Noah and I spotted more queens as the season unfolded, and then we began seeing workers, too, proof that nests were getting started somewhere nearby. Our first attempt to find one began with a quick trip to the chicken coop.

In our small flock, the longest-surviving hen is a large Buff Rock named Golden. She'd begun putting on weight in her older years, which made for a tight fit squeezing in and out of the coop's narrow door. The result was a handy supply of loose feathers lying around—fluffy yellow ones that seemed like they'd be pretty easy to follow, trailing behind a bumblebee. We selected a nice fresh plume, trimmed it to size, and then returned to the house, where Noah promptly caught a bumblebee on the flowers of a nearby gooseberry bush. After chilling our subject briefly in the freezer (a recommended tactic for pacifying any cold-blooded creature), I dabbed the top of the bee's abdomen with water-soluble glue and applied the feather. Then we placed her on the top step of the porch and crouched nearby, boots on, ready to give chase.

It took a moment for her metabolism to warm back up to speed, but soon she was busily grooming her antennae and looked ready for flight. We watched as she pumped her abdomen and shivered, distributing muscle-generated heat throughout her body. Then, with what begs to be described as an irritated motion, she reached up suddenly with one back leg, grabbed the feather, and yanked it off.

In a popular book of children's verse, the nineteenth-century English poet Sarah Coleridge once mused: "I wish we could feel, though it were but in part; / The kindness that glows in a Bumble Bee's heart." Ms. Coleridge had clearly never tried gluing a feather to her bumblebee. She might have penned a different couplet if she'd seen our bee savagely wadding the offending plume into a gluey ball with all six of her feet before stalking off to a patch of sunlight, buzzing her wings briefly, and flying away out of sight. We tried several variations on this theme with similar results. Bumblebees may look like clumsy teddy bears, but their legs are extremely nimble, adapted for reaching up to groom pollen from virtually anywhere on their bodies. Even the smallest, stickiest plumes were quickly dealt with, and it turns out that a bee can also remove a feather tied on with thread. Taking a different tack, we powdered a few bees with bright blue chalk dust, which made them stand out nicely against leaves or lawn. (The carpenter bees of Malaysia, naturally vibrant with thick blue fuzz, must be a cinch to track as they wing their way through the rainforest.) Unfortunately, our chalky bees disappeared completely whenever they crossed a backdrop of blue sky, leaving us no more than a few running steps closer to their hidden nests.

What worked in the end was something that honey hunters probably take for granted, a heightened and habitual awareness of bees that amounted to a constant search. Our heads began to turn at every passing buzz, and we started paying particular attention to what Noah aptly called "suspicious bees"—a queen investigating the roots of an upturned stump, or workers hanging around in places with no obvious sources of pollen or nectar. When I saw a bee fly out of an old horse shed near the garden, it didn't take us long to discover not one, but two nests inside. A Sitka bumblebee queen had set up housekeeping beneath an old wooden pallet, less than ten feet from an abandoned vole tunnel inhabited by a species known as the fuzzy-horned bumblebee. With a deck chair placed halfway in between, I realized that I could observe the activity at both nest entrances *and* work on this book at the same time. It turned out to be a productive place to write. Out of reach of phone and email, my only interruptions took the form of the pleasant comings and goings of bumblebees.

At first I saw just the two queens, doggedly heading out on flight after flight and returning with pollen packed tight to their hind legs. During those first, critical weeks of colony development, the queen does everything—gathering provisions and laying eggs much like a solitary bee. But if I could have peered inside I would have seen that these nests were clearly different from those of masons, diggers, or alkali bees. Instead of sealing her eggs away in separate chambers, a bumblebee queen lays them in groups and broods them like a bird, using her body's warmth to speed their development. From the perspective of my chair, I could guess what each queen was doing simply by looking at the clock. Dropping a load of pollen or nectar in the nest took as little as a minute, but when there were eggs to brood she might stay inside for close to an hour between foraging flights. It became like a race to see which nest would hatch out the first workers, but when that moment finally came, I almost missed it completely.

"Tiny!" my notes read, describing the two black insects I spotted buzzing up from the old pallet that hid the Sitka nest. They looked like houseflies, but with little tufts of whitish hair decorating their abdomens. In the parlance of bumblebee studies, these newly

emerged workers were known as *callows*, and their size was a simple reflection of their limited diet. Raising her first brood alone, a
queen often can't supply them with enough pollen to grow to their
full potential. In a sense, she sacrifices size for expedience, eager to
establish the caste-like division of labor that defines the life of a
social bee colony. Eventually, nurse bees, guards, and a legion of
other workers would take on the tasks of maintaining a growing colony, allowing the queen to devote herself to laying eggs. With more
and more adults gathering pollen and tending the larvae, bees from
future broods would be up to ten times larger than those the queen
raised on her own. I knew that all of this was underway now that I
spotted those first two callows, but I still had mixed feelings when
I watched them heading out to forage. It meant that I had probably
seen their big, lumbering queen for the last time. The same division
of labor that put workers in charge of gathering pollen and nectar
meant that the queen would spend the rest of her life in the darkness of the nest, an egg-laying machine surrounded by her children
in an ever-growing network of brood cells, pollen stores, and honeypots. As it turned out, however, I never saw the queen, the callows,
or any other bees from that colony again. By the next time I settled
in to my deck chair for another round of observations, the Sitka
bumblebee nest had gone silent.

Charles Darwin once linked the fate of certain wildflowers to the
prevalence of house cats, noting that cats eat mice, that mice eat
bumblebee nests, and that bumblebees are essential pollinators for
things such as red clover and heartsease, a variety of wild violet. He
concluded: "Hence it is quite credible that the presence of a feline
animal in large numbers in a district might determine, through the
intervention first of mice and then of bees, the frequency of certain
flowers in that district!"

Later commentators expanded this model to include spinsters in
rural English villages (who often kept cats), and sailors in the Royal

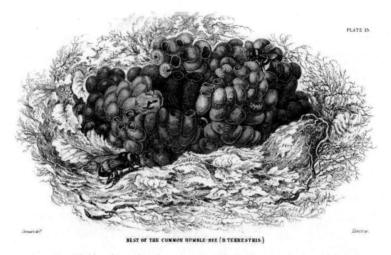

PLATE 19.

NEST OF THE COMMON HUMBLE-BEE (B TERRESTRIS)

FIGURE 7.3. Unlike the ordered symmetry of honeybee comb, bumblebees store their provisions and brood in a haphazard collection of tiny wax pots. WIKIMEDIA COMMONS.

Navy (who ate salted beef from cows pastured on clover), therefore tying the defense of the British Empire to the number of its cat-loving spinsters. This anecdote is often held up as an amusing early example of the food chain concept, but on Darwin's part it also reveals a keen understanding of bumblebees. Sladen, Plath, and other authorities all confirmed that rodents do indeed prey on bumblebee colonies, particularly new ones like my Sitka nest, where only a few small workers were available for defense. I could think of no better explanation for what I found when I lifted the pallet and picked through what remained below. After all, I knew that there were rodents living in the shed and the field outside, and that they had probably been the first inhabitants of the nest. It consisted of two small chambers tunneled into a tangled mass of dried grass, poplar leaves, baling twine, bits of fabric, and shreds of foil from a shiny granola bar wrapper. There was no sign that an animal had torn into it from the outside, and no evidence of dead or diseased bees within. Instead, some curious mouse or rat had probably just followed the tunnel to the front door, overpowered the callows, and devoured everything in sight. The only sign that bees had ever called the place home was part of a single, urn-shaped capsule of tawny wax.

I was sorry to lose the Sitka nest, but its demise drove home something I'd failed to realize, in spite of all the false starts that Noah and I experienced with our various boots and contrivances. (In addition to the attempts already described, we had also lost a nest box to raccoons, and seen a queen's tentative first steps overrun by ants.) As anyone caring for honeybee hives might have told us from the start, keeping bees is hard work. To establish and maintain a healthy colony requires overcoming a succession of natural obstacles, from pesky competitors to the vagaries of weather to the constant threats of predators, parasites, and disease. Even in the wild, success is the exception rather than the rule. If it were otherwise, and every queen founded a thriving colony, the result would be an unsustainable glut of bees. I took solace in what Noah and I did gain from our efforts, an affinity for bumblebees that continues to help us locate wild nests, everywhere from the woods around our house to the cracks in city sidewalks. And I still got to watch the fuzzy-horned nest in the vole tunnel, which proliferated, becoming so busy with workers that I could no longer pretend to be writing as I watched them flying in and out. As the weeks passed, their pollen loads tracked the progress of our nearby garden, changing from the bright orange they picked up at the asparagus plants to the black dust of poppies and the whitish grains from squash and melon. It was a fitting way to conclude any attempt at beekeeping, because no matter how fascinating their biology or how much we enjoy their honey and wax, our deepest connection to bees comes in how they affect what we eat.

CHAPTER EIGHT

Every Third Bite

Tell me what you eat, and I will tell you who you are.

—French Proverb

It's often said that every third bite of food in the human diet relies upon bees. For a Hadza hunter during the peak of honey season, that figure may be an underestimate. For the rest of us, it alludes to the great debt we owe bees for pollination, a largely unheralded service at the heart of our agricultural system. Parsing the numbers to reach "every third bite," however, can be challenging. Measured by volume, 35 percent of global crop production comes from plants that depend on bees and other pollinators. That's pretty close to one in three, but doesn't take into account all the calories we get from meat, seafood, dairy, or eggs. In terms of simple food variety, the ratio looks more like three out of four: over 75 percent of our top 115 crops require or benefit from pollinators. Nutritionists take a different approach, pointing out that pollinator-dependent fruits, vegetables, and nuts provide over 90 percent of our vitamin C, as well as all of our lycopene and the vast majority of our vitamin A, calcium, folic acid, lipids, various antioxidants, and fluoride.

Pollination clearly makes a big impact on our food, but the importance of bees to any particular bite depends on what you're biting into. Cows and other edible animals can be raised without pollinators, and staples like wheat and rice come from wind-pollinated grasses. If you want to add flavor to your meat, however, or spread something tasty onto your bread, things quickly get more complicated. Rather than focusing on how bees impact food quantity, it might be more revealing to examine their effect on quality. We could still find things to eat in a world without bees, but what would our food be like? Visiting a produce aisle or farmers market would certainly be different, the selection reduced from colorful profusion to a few grains, a nut or two, and oddball clones like bananas. (Even reliable self-pollinators like peas or eggplant were originally developed from bee-pollinated strains.) But that's the obvious change— less choice in fruits and vegetables. To really see the pervasiveness of bees in our food supply, I decided to look for them someplace totally unexpected and unlikely, in a meal served over two and a half million times every day in more than one hundred countries around the world. Its ingredients are straightforward and at first glance seem far removed from the influence of buzzing insects. I know this because, like millions of other people, I happen to be able to sing the recipe.

Introduced at a Pennsylvania McDonald's franchise in 1967, the Big Mac sandwich was added to menus nationally a few years later. But it didn't become a sensation until 1975, when the company debuted one of the most successful advertising jingles of all time: "Two all-beef patties, special sauce, lettuce, cheese, pickles, onions—on a sesame seed bun!" For a limited time, customers who could blurt out the whole phrase in less than three seconds were given the burger for free. Though I hadn't eaten one since high school, I remembered the flavor well and began to wonder what, if anything, bees had to do with it.

Living on a rural island offers the advantages of clean air, bird-song in the morning, and ready supplies of firewood. But reaching a pair of bona fide "Golden Arches" in time for lunch required me to leave the house before I'd even finished digesting breakfast. After an hour and a half on a ferryboat, followed by a brisk bicycle ride into the nearest town, I arrived at McDonald's hungry enough to risk de-vouring my Big Mac before I'd even examined it. As I waited in line I could hear the din of fryer alarms and oven timers ringing from the kitchen, where a row of people stood shoulder to shoulder, assem-bling and wrapping burgers with lightning speed. I tried to watch them build my order, but it was no use—their hands were a blur.

For those who've never had one, a Big Mac sandwich comes with three layers of bun and two layers of meat, all gooped up with sauce and onions. The pickles lie below the top meat patty, and the cheese goes underneath the lower one, where it melts slightly and droops down over the bottommost section of bun. Handfuls of shredded let-tuce and chopped onions get sprinkled in with the sauce, tucked un-derneath each meat patty. Armed with tweezers and a hand lens, I began disassembling this construction, layer by layer, and removing any ingredients that wouldn't be available without the assistance of bees. (For reference purposes, I'd also brought along a detailed list of ingredients and nutritional information printed from the McDon-ald's corporate website.) Here are my results, in the order laid down by that famous advertisement.

The two all-beef patties could stay. McDonald's sources its meat from several major distributors that, in turn, buy from thou-sands of farms and cattle ranches. Some of those cows probably did nibble on a bit of bee-pollinated alfalfa or clover, and feedlots have been known to fatten up their charges with all manner of food-industry cast-offs, from surplus ice cream sprinkles and gummy worms to bee-pollinated cherry juice and fruit fillings. But with

few exceptions, the vast majority of a beef cow's diet comes from
wind-pollinated grasses and grains. In terms of seasoning, McDonald's adds salt to their meat, which is fine, but they also sprinkle it
with pepper, which raised the first potential red flag. Black pepper
comes to us from a tropical vine in the genus *Piper*, native to southern India. Stingless bees visit its flowers regularly, but many pepper
varieties are self-fertile, and some experiments suggest that wind, or
even the jostling of raindrops, can distribute enough pollen to set a
good crop. Since the flecks were too small to remove anyway, I decided the pepper could remain.

Not so, the special sauce. A variant of Thousand Island dress-
ing, this creamy, pinkish condiment includes a sweet pickle relish
made from bee-pollinated cucumbers, as well as a powdered form
of onion, a bulb crop that requires bees for seed production and
breeding new varieties. The sauce gets its color from paprika, a bee-
pollinated pepper, and turmeric, from the root of a bee-pollinated
herb in the ginger family. Its creaminess comes from either soybean
oil or canola oil. While soybeans can self-pollinate, the assistance
of bees improves their yield by anywhere from 15 to 50 percent.
Canola—a trade name for field mustard—also depends on bees for
healthy yields, as well as for the production of viable seed. Without
bees, then, the only things left in the sauce would have been corn
syrup, egg yolks, preservatives, and minor ingredients with names
like "propylene glycol alginate" (a thickener derived from kelp).

In removing the dollops of special sauce, I ended up scraping off
most of the lettuce, too, which was probably just as well. Though
we only eat its leaves, and the plant can produce seed from self-
pollination, sweat bees and other species do visit lettuce flowers,
improving fertilization rates dramatically and transporting pollen
between plants as far as 130 feet (40 meters) apart. What's more,
the crispy lettuce preferred by McDonald's would never have arisen

without the assistance of bees. Famed seed man Washington Atlee Burpee developed the "iceberg" lettuce variety in the early 1890s, during a series of open-pollinated trials at his farm in Pennsylvania.

As another product of cows, the slice of cheese on the Big Mac at first appeared to be a safe, bee-free bet. But while beef cattle eat mostly grass and grains, a bit of research told me that dairy cows scarf up the vast majority of the world's alfalfa, which I knew from experience depended on alkali bees and leafcutters. With its high protein and mineral content, alfalfa makes ideal fodder for milk production, and industry guidelines suggest daily rations of fourteen to sixteen pounds of the stuff for every lactating member of a herd. Those cows could of course survive on grass alone, but the resulting dairy products would be less plentiful and more costly, and might not find a place on an inexpensive fast-food burger. The point was debatable, but alfalfa wasn't the only way that bees impacted the cheese slice. It also included an emulsifier derived from soybeans, and it got its distinctive yellow color from the bright seeds of annatto, a tropical tree pollinated by various South American bumblebees. I therefore peeled it off, as well as the more obviously bee-related pickles and onions. That left only the bun, for which my information from McDonald's listed fifteen ingredients in addition to wheat flour. Like the flour, the other ingredients were mostly bee-free, or had simple bee-free replacements, with the exception of the sesame seeds. As one of the world's oldest cultivated plants, sesame was selectively bred long ago to produce self-fertile varieties. No one has studied its biology in cultivation, but photographs of its showy, zygomorphic flowers leave no doubt that it began life like its wild relations, pollinated almost exclusively by bees. Using the tweezers, and with more than a few curious glances from the family at the next table, I removed all 243 sesame seeds from the top of the bun and put them in the discard pile.

FIGURE 8.1. A Big Mac deconstructed, with the relatively bee-free meat patties and bun on the left, and all the bee-reliant ingredients, from pickles to special sauce to sesame seeds, arranged on the right. PHOTO © THOR HANSON.

Deprived of its bee ingredients, my Big Mac now looked rather sad and unappetizing. In this form, it's hard to imagine it would ever have become the world's most popular burger. Certainly, the advertising slogan wouldn't have been nearly as catchy: "Two all-beef patties, bun." Like the Big Mac, almost any meal can be deconstructed and examined for the influence of bees. Try it, and you'll learn what I learned: yes, we could still eat in a world deprived of its primary pollinators, but eating would be extremely dull (and not very nutritious). As I picked at the remains of my lunch, I realized I couldn't even console myself with an order of french fries. McDonald's uses a potato called the Russet Burbank, developed from the seeds of an open-pollinated Early Rose variety by the celebrated plant breeder Luther Burbank (Washington Atlee Burpee's cousin). Sprucing

TABLE 8.1. A list of 150 crops that either require or benefit measurably from pollination by bees. Some rely completely on bees to produce fruit or seeds, while others show increased yields when bees are present. ADAPTED FROM MCGREGOR 1976, ROUBIK 1995, BUCHMANN AND NABHAN 1997, SLAA ET AL. 2006, AND KLEIN ET AL. 2007.

Alfalfa	Chili pepper	Lemon	Prickly pear
Allspice	Chives	Lentil	Pumpkin
Almond	Citron	Lettuce	Quince
Anise	Cloudberry	Lime	Radicchio
Annatto	Clover	Loquat	Radish
Apple	Cloves	Lychee	Rambutan
Apricot	Coconut	Macadamia	Rapeseed
Artichoke	Coffee	Mandarin	Raspberry
Asparagus	Collards	Mango	Red currant
Avocado	Coriander	Marjoram	Red pepper
Barbados cherry	Cotton	Medlar	Rose hips
Basil	Cowpea	Millet	Rosemary
Bay leaf	Cranberry	Muscadine grape	Rowanberry
Beans (various)	Cucumber	Muskmelon	Rutabaga
Bergamot	Cumin	Mustard	Safflower
Black currant	Dewberry	Nectarine	Sage
Blackberry	Dill	Nutmeg	Sapote
Blueberry	Durian	Oil palm	Sesame
Brazil nut	Eggplant	Okra	Soybean
Breadfruit	Elderberry	Onion	Squash
Broccoli	Endive	Orange	Starfruit
Brussels sprouts	Fennel	Oregano	Stevia
Buckwheat	Fenugreek	Papaya	Strawberry
Cabbage	Flaxseed	Paprika	Sugarcane
Canola	Garlic	Parsley	Sunflower
Cantaloupe	Grapefruit	Parsnip	Sweet potato
Caraway	Groundnut	Passionfruit	Tamarind
Cardamom	Guar	Peach	Tangerine
Carrot	Guava	Peanut	Thyme
Cashew	Hog plum	Pear	Tomatillo
Cassava	Jackfruit	Pepper	Tomato
Cauliflower	Jujube	Persimmon	Turnip
Celeriac	Kale	Pigeon pea	Vanilla
Celery	Kiwifruit	Pimento	Watermelon
Chayote	Kohlrabi	Plum	Yams
Cherry	Kola nut	Pomegranate	Zucchini
Chestnut	Kumquat	Pomelo	
Chickpea	Leek	Potato	

things up with bee-dependent mustard or tomato ketchup was, of course, out of the question. In the end, I did with my Big Mac what we would all have to do in a bee-free world: I ate what I could.

Whether measured by quantity, variety, nutrition, or flavor, nearly every bite of food we take feels some effect from bees. But it's worth pointing out that other animal pollination options do exist. Flies, wasps, thrips, birds, beetles, and bats do a bit of crop pollina- tion, and, in a pinch, so do people. Gregor Mendel hand-pollinated over ten thousand pea plants in his pioneering study of genetics, and modern plant breeders use similar techniques to create new hy- brids or to cross particularly promising varieties. But for anything produced on a commercial scale, pollinating by hand is usually con- sidered too labor intensive to be anything more than a last resort. One notable exception to that rule comes in the form of a sweet, hot-country fruit once considered sacred from Egypt to Babylon. Now cultivated in deserts around the world, its annual harvest re- cently topped seven and a half million metric tons—more than avocados, cherries, and raspberries combined. Getting all that fruit pollinated requires its growers to become, for a few weeks every year, the equivalent of human bees. Almost no other crop in cultivation requires such an effort, and nothing better explains the debt we owe to bees than seeing that process in action.

When I met Brian Brown, he was chewing on a date. "I still eat them," he said, and then grinned as if a little surprised at himself. For someone who'd spent more than thirty years planting and tend- ing an orchard of date palms that now totaled more than 1,500 trees, maybe it was a little surprising. I watched as he spit the seed into his hand with a practiced motion and tossed it into a nearby urn labeled "Pittoon." Then he turned back to me. "Now, what was it you wanted to see?"

We were standing outside the café and gift shop at the China Ranch Date Farm, an oasis of green in the middle of California's Mojave Desert, just a few miles from the entrance to Death Valley. When I reminded Brian of our email exchange, his eyes lit up. "Pollination, right!" he said, and led me quickly into a back room to gather supplies. Soon we were bouncing across a field in his pickup truck, armed with cotton balls, a wad of twine, and a wicked-looking curved knife.

"These here are Khadrawy, an Iraqi variety," he said as we braked to a stop in the midst of a date grove. Then he was out of the truck and propping an aluminum extension ladder up against the side of a tree, wedging its top between the bases of two spiky green fronds. "Luckily, they've already been de-thorned," he said, and explained how the first step in hand-pollination involved cutting off rows of needle-sharp, six-inch spines from the bases of all the surrounding leaves. (This topic came up again later, when we were discussing workers' compensation insurance. "It's a job with heights, spines, and knives," he said, shaking his head. "My rates are through the roof.") Once the ladder was secure, Brian looped some lengths of twine through his belt, grabbed the jar of cotton balls, and tucked the knife into the back pocket of his jeans. Then, with the ease of long practice, he strode up the ladder as if crossing level ground and scrambled into the crown of the palm.

"You can use pollen from any male tree," he called down. "The fruit stays true to the female." With that he summed up a key biological fact about date palms. They are what botanists call *dioecious*, from the Greek phrase for "two houses," which means that individual trees are either exclusively male, producing pendant six-foot flower clusters laden with pollen, or female, like the one that Brian was climbing. "We thin about a third of the flowers or the fruit would be too small," he said, shaking loose a few yellowish strands from the nearest mass. I picked one up from the ground nearby and saw that its entire two-foot length was studded with tiny flower nubs, each one a potential date in the making.

Left to their own devices, the trees rely on wind for pollination. But while that may be a successful strategy for conifers, grasses, and many other plants, the process seems imperfect in date palms, or at least too erratic to produce a reliable crop. Even in a well-organized orchard, most female flowers would wither before ever receiving pollen from wind alone. For at least four thousand years, growers have known that hand-pollinating dates is the only way to make them commercially viable, improving yields by a factor of five. The Egyptians did it, as did the Assyrians, the Hittites, the Persians, and virtually every other culture from North Africa and the Middle East.

FIGURE 8.2. The human equivalent of a bee: orchardist Brian Brown hand-pollinating a date palm at his China Ranch Date Farm in California's Mojave Desert. PHOTO © THOR HANSON.

Passed down from generation to generation, their pollination expertise transformed dates from a seasonal wayside snack into the staple
fruit of the ancient world.

Watching Brian at work in his tree, I reflected on how little the
process had changed since Greek scholar Theophrastus described it
in the third century BCE: "When the male palm is in flower, they at
once cut off the spathe . . . and shake the bloom with the flower and
the dust over the fruit of the female." Instead of using a whole flower
stalk, however, Brian doused the female bloom with a pollen-laden
cotton ball from the jar, running it up and down each strand to
make sure it touched every flower. "Then we tie the flowers around
the cotton," he called down, as he deftly pulled two strands of twine
from his belt and wrapped them around the long inflorescence.
Leaving the cotton in place allowed more pollen to shake out over
time, fertilizing any late bloomers. Ultimately, there would be some
wind pollination too from the male trees scattered throughout the
orchard. But their flowers hadn't opened yet, and Brian was starting
the season with last year's pollen, which had over-wintered just fine
in the café's large freezer, right next to the ice cream they used to
make their signature date-whip milkshakes.

Before we moved on, Brian carefully showed me the whole process again, stopping at each stage so that I could ask questions or take
pictures. It dawned on me that it wasn't the first time he'd taught
someone how to pollinate a date palm. "Actually, I just trained two
people this morning," he admitted, and our conversation shifted to
the challenges of staffing. In any given season, his pollination crew
consisted of himself and various full- and part-time local employees,
supplemented by volunteers from around the world. They came on
working vacations, learning the date trade in exchange for room
and board. "It's like internet dating for business," he explained.
His bunkhouse already held boarders from Belgium, Germany, and
Montreal, and more were expected any minute. "I've got a couple

from Russia coming in today," he said, "and a whole family from France arriving tomorrow." As my tour of the China Ranch Date Farm continued, it became clear that Brian would have no trouble keeping them all busy.

"The flowers open sequentially," he said, and explained how a healthy tree produced from ten to twenty clusters that would mature into drooping bunches of dates weighing as much as seventy-five pounds (thirty-four kilograms) each. But they bloomed unpredictably, requiring daily inspections of every tree to catch the opening clusters at just the right moment for pollination. The multiple climbs that followed might involve a ladder, a tractor with a raised platform, or, for the oldest and tallest palms, the type of boom lift more typically used to reach telephone wires and stadium lights. Male trees required monitoring, too, and repeated harvesting of flowers, which had to be dried and rubbed across window screens to extract their pollen. "The whole crop is labor," Brian observed at one point. He reminded me that both harvesting and processing the dates also involved intensive handwork, and that the fruit had to be protected, too—from birds as well as a more surprising pest. "Coyotes love dates," he said, with a resigned shrug. "They pick them off the lowest trees—they'll stand right up on their hind legs to reach them!"

We ended our tour in the driveway beside Brian's home, a low rambler that he and his late wife had constructed from eighteen thousand mud bricks, hand-made on site. It occurred to me that date farming was a particularly good fit for Brian—he seemed to like doing things the hard way. "I'm an unorthodox guy," he acknowledged. Apart from a few years studying agriculture at Colorado State University, he said, he'd spent his whole life in the vicinity of China Ranch. With tanned features and a perpetual, blue-eyed squint, he certainly looked like someone very much at home in the desert. As if to illustrate that point, he stopped in mid-sentence when a bird called from the parched hillside behind the house: four hollow, descending notes, like a dove blowing in a bottle. "Do you hear that mournful cry?" he asked quietly. "That's a male roadrunner trying to call in a girlfriend." He paused for a moment, and then continued telling me the history of the business—how he and his wife had transplanted unusual date varieties from abandoned orchards around the Southwest, and made their first sales from the tailgate of their truck. We walked over to inspect an old grove of male palms shading the house. The bird on the hill had gone silent, but I was pretty sure that no matter what we were talking about, Brian was still listening for it.

While a deconstructed Big Mac tells us what food might be like without bees, the date palm suggests the toil and strain that would be required to replace them. Pollinating a midsized operation like China Ranch involves more than six thousand trips up and down the palm trees. That's a quantity of labor that other fruit growers get largely for free, from bees. Even if they rent the services of commercial hives, their expenses come nowhere near what date farmers pay to have the job done by people. Bees, for example, do not require workers' compensation insurance. Again and again, Brian stressed the importance of good pollination in making his business viable, but when I pressed him on how much it added to the cost of production, he demurred. "I wouldn't want to do that calculation," he said. "It would be too depressing." I found an answer, however, by simply visiting my local grocery store, where California dates were selling for $9.99 per pound, more than twice as much as anything else in the produce section. If I'd wanted to spend more on a piece of fruit, my best option lay in a trip to the spice aisle. There, I could have plunked down $27.50 for a pair of the orchid pods known as vanilla beans. Perhaps not surprisingly, vanilla is the only other major crop in the world that relies primarily on hand-pollination.

After wrapping up my conversation with Brian, I decided that no visit to China Ranch would be complete without one of the gift shop's famous date shakes. To work up the proper thirst, I took a hike through the surrounding desert, making my way down to a little slot canyon on the edge of the nearby Amargosa River. The trail took me past leaning stone walls from an abandoned homestead, and a pile of white tailings where someone had mined for gypsum. Creosote bush and clumps of low cacti stretched in all directions, and the surrounding hills looked sun-blasted and raw, like bedrock crumpled up and tossed aside by giants. It was a far cry from the evergreen forests I was used to, but there was a grandeur and vast silence to the place, and I could see small beauties that a person might fall in love with. While I'd timed my trip for the early spring date bloom, I saw now that the first desert wildflowers were also coming to life—unexpected splashes of yellow sunflower, and the occasional vivid blues of phacelia. Hoping for bees, I found a good patch and settled down to watch.

Quiet minutes passed without any sign of pollinators until I noticed the cheerful flit and dither of a western pygmy blue, the smallest butterfly in North America. With a wingspan of less than half an inch (twelve millimeters), it seemed a lonely speck to be responsible for so many flowers. But no bees turned up to help. Logically, I knew that I was probably just too early in the season. The habitat looked perfect, and there must have been droves of dormant bees all around me, tucked into their winter homes in ground nests, cut banks, rodent holes, or the hollow ends of twigs and stems. As temperatures warmed and the bloom expanded in the days ahead, those bees would surely emerge and buzz the desert back to life. I knew all of that, but still felt bothered long after I'd finished my hike, enjoyed my milkshake, and bid China Ranch goodbye.

In the twenty-first century, the absence of bees can't always be written off as a quirk of timing. While I was at work on this book, a group of more than eighty bee experts from around the world published the first global assessment of pollinator populations.

Wherever data were available on bees, they found that roughly 40 percent of species were considered to be in decline or threatened with extinction. The finding made headlines—suddenly, discussions of a world without bees seemed like more than just a thought experiment. In the chapters ahead, this narrative will shift from stories about the biology of bees, and our connections to them, to a frank look at their prospects. It begins in the field, with someone whose deep scientific experience might be matched, or even exceeded, by his capacity for hope.

The Future
of Bees

To make a prairie it takes a clover and one bee,
One clover, and a bee.
And revery.
The revery alone will do,
If bees are few.

—Emily Dickinson, undated

CHAPTER NINE

Empty Nests

The important thing is to not stop questioning.
—Albert Einstein,
"Old Man's Advice to Youth" (1955)

The meadow stretched seductively across a small mountain basin fringed by oak, fir, and ponderosa. From the edge I could see dozens of different wildflowers in full bloom—spires of purple lupine towering over a riot of asters, geraniums, saxifrage, and vetch. I'd spent eighteen hours on the road for this moment, to stand in perfect bumblebee habitat alongside one of the world's leading bumblebee experts. There was only one problem.

"Too bad about the weather," said Robbin Thorp.

Overhead, storm clouds eddied and flowed in a low, dark mass. An icy wind swept down from the hills, and I wished that I'd thought to pack a winter jacket. My fingers, gripping the wooden handle of my bee net, already felt numb.

But Robbin seemed unperturbed. In a career spanning more than six decades, he'd learned how to make the most of any field day.

Wearing a floppy sun-hat and tinted glasses, with his snow-white beard cropped short, he looked something like Santa Claus on vacation—assuming the old elf spent his off-season in California, doing a lot of hiking. "Let's see what we can find on the flowers," Robbin said as we set out. "Look closely at the balsamroot—they love to sleep on that."

Crossing a fence line, we slowly made our way through the tall grass, bending now and then to inspect a blossom, and always listening for the buzz of wings. A young student from one of Robbin's bee workshops, Langdon Eldridge, had come along to get some field experience. After a few moments, he called out with the first find.

We hustled over, and there, clinging to the underside of a geranium petal, was a big black and yellow bee. To my surprise, Robbin reached out with what looked like a large plastic water pistol and pulled the trigger. A tiny motor whirred to life, and the bee disappeared up the barrel. "These things work great," he said, and showed me the name emblazoned on the side: "Backyard Safari Bug Vacuum." (The company's promotional materials note that "Kids love catching bugs!" But apparently they also do a brisk business with entomologists.) The bee had dropped into a transparent "Capture Core" for easy viewing, but Robbin immediately shook it out onto his palm.

"Definitely queen-sized," he said, staring down at the unmoving form. The bee appeared to be stiff with cold, or perhaps she was still asleep. "She should warm up in a minute," he went on, and then pointed out her distinctive features: dense black hair on the face, and a fuzzy black abdomen striped with a single band of yellow. Langdon correctly identified the species as *Bombus californicus*, the California bumblebee, and Robbin seemed pleased. After that we were quiet for a moment while he tipped the motionless insect back and forth on his palm. Finally, he admitted, "I think she's dead."

It was impossible to know exactly what had killed the bee. Perhaps she'd been attacked by a crab spider, or, considering the snowflakes that had begun drifting down around us, maybe she'd just gotten too cold. Either way, it wasn't exactly an auspicious beginning to our bumblebee search. But I suppose things could have been worse. After all, by most accounts, the particular species we were trying to find had already gone extinct.

"I had no idea that I would be witnessing a devastating decline," Robbin told me, recalling the day he had taken on a small consulting project for the United States Forest Service. It was in the late 1990s, and they wanted him to look for rare bees in Oregon's Rogue River Valley, a flashpoint in the controversy then raging about spotted owls and the logging of old growth forests. The agency had decided to look more broadly at the ecosystem, rather than just focus on one species. "If there were other special things in the area," Robbin explained, "they thought it might help take some of the pressure off the owl."

Robbin's target was Franklin's bumblebee, a little-known species found only in southwestern Oregon and neighboring parts of California. It looked similar to the California bumblebee, but sported bright yellow shoulders and a yellow face. He'd seen it before, in the wild and as a pinned specimen in various collections. With scores of scientific articles under his belt, as well as monographs and books bearing titles like *Bumblebees of North America*, there weren't many species in the genus *Bombus* that Robbin couldn't tell at a glance. Armed with a list of locations where the bee had been spotted in the past, he left his office at the University of California in Davis and headed for the Rogue Valley.

"I found it at all of its historical sites in 1998," he remembered. "It wasn't the most common bee, but it was there." The following year was much the same, though he had to look harder to find it.

Then it simply dropped off the map. In 2000, Robbin located only nine individuals, and in 2003 he saw fewer than five. By that time he'd widened his search to sites throughout the bee's known range, and alerted colleagues that something drastic was going on. Local biologists kept watch and the federal Bureau of Land Management sent out a survey team, but no one could find a trace of it. In 2006, Robbin spotted a single Franklin's bumblebee, a worker, foraging on buckwheat blossoms in a subalpine meadow. Nobody has seen one since.

"I keep hoping it's still out there, flying under the radar," Robbin told me at one point. We had crossed the field and started up a rise on the opposite side, where the grasses and wildflowers extended into gaps between the trees. The snow had stopped, and there were now a few bumblebees braving the chilly air. We'd seen no sign of the elusive Franklin's, but that doesn't necessarily mean it wasn't there. In biology, it's often difficult to prove a negative, particularly the absence of something tiny and hard to find. Robbin said it wasn't unusual for a small population of insects to persist undetected for a long time. "If I keep looking, there's still a chance it will turn up," he said, and then glanced over at Langdon, who was ranging far ahead of us up the slope. "And if I train more people to look, they'll get into a lot of areas I'd never get to."

It remains to be seen whether Robbin Thorp has witnessed an extinction or just a steep decline, but one thing is certain: the Franklin's bumblebee couldn't have found a better champion. Since that last sighting in 2006, Robbin has doggedly continued his annual monitoring, patiently scouring the meadows and roadside flowers of southwestern Oregon year after year. Some people find his efforts quixotic, and he's become a bit of a celebrity. He was once featured in a CNN segment called "The Old Man and the Bee." But while others might have given up long ago, and Robbin's quest remains unfulfilled, his hundreds of hours in the field did put him in a prime position to notice something else. Franklin's bumblebee was not the only species in trouble.

"It took a couple more years before I realized it was a trend for *Bombus occidentalis*, too," Robbin explained, referring to the western bumblebee. Unlike the Franklin's bee, which had always been rare, the western was, until recently, among the most prolific bumblebees west of the Rocky Mountains, from Mexico north to Alaska. (It was so common that researchers once assumed it was the species that my cliff-dwelling digger bees had evolved to mimic.) But shortly after he stopped finding Franklin's bumblebees, the western dropped out, too, disappearing not only from Robbin's surveys but from much of its former range. At the same time, entomologists in eastern North America began sounding the alarm about two other once-common species, the yellow-banded and rusty-patched bumblebees. It became apparent that Robbin, who had spent the first part of his career as a student of bees, would need to devote the remainder of it to a new role: bee detective.

"My theory was that it must be a pathogen," Robbin told me. "Other bumblebees were doing perfectly well in the same habitat," he went on, which appeared to rule out pesticides or some kind of disturbance. Then he explained how all four of the dwindling species were closely related, part of what taxonomists call a subgenus. That made them potentially susceptible to the same strains of various afflictions, from viruses to fungi, mites, bacteria, or parasites. But while at first he didn't have a clear idea of what the pathogen might be, he did have a strong suspicion about its source, an enterprise that helps keep one of the world's most popular fruits in season all year long.

Tomatoes were domesticated in ancient Mexico, Central America, or possibly Peru, and nobody knows who grew the first one, but the history of greenhouses is more clear-cut. Credit for the first such building goes to a group of gardeners employed by the Roman

emperor Tiberius in the early first century. With a roof of translucent minerals such as mica and selenite, the structure allowed
year-round production of the emperor's favorite melon, a relative
of the modern cantaloupe. As Pliny the Elder recalled, "there was
never a day on which he was not supplied with it." Greenhouses
remained an indulgence of the rich, however, until the Industrial
Revolution provided enough inexpensive glass (and later plastics)
to make them economically feasible on a large scale. Early commercial ventures produced a range of fruits, vegetables, and flowers, but
one crop quickly became established as the most prolific and profitable greenhouse product in Europe: tomatoes. Cultivation methods
became increasingly sophisticated, particularly in the years following World War II, ensuring consistent year-round yields in countries
as far north as Belgium, the Netherlands, and the United Kingdom.
The idea didn't catch on until much later in North America, where
traditional tomato farming had always benefited from the long, hot
growing seasons in places like Florida and California. When demand
for greenhouse varieties finally started to rise in the 1990s, Canadian
and American growers immediately looked to their European counterparts for advice. And one of the first things they learned was an
unexpected axiom in the tomato business—unless you want to buy a
lot of electric toothbrushes, you're going to need some bumblebees.

The connection between bees and brushes boils down to their
buzz. If you've never tried one, I can tell you that using an electric
toothbrush feels something like chewing on a tuning fork. The model
I use vibrates with a penetrating hum pitched at high C, which my
dentist assures me does a bang-up job on plaque. But it's also similar
to the tone that bumblebee wings make during the remarkable process of buzz pollination. Watch them visit a tomato (or some other
buzz-pollinated species, like eggplant or blueberry), and you can see
this in action, or at least hear it—a quick, high-pitched buzz emit-
ted every time the bee lands on a flower. Like other members of

the nightshade family, tomatoes have what botanists call "poricidal anthers," a design that holds their pollen in tiny chambers accessed only by a small hole (the pore) at one end. While some shakes out naturally over time, allowing a certain amount of self-fertilization, vibrations at the right frequency make the anthers resonate and release a spray of pollen through the pore. From the plant's perspective, this strategy creates a special bond with the few pollinators, like bumblebees, that have figured out the trick. Honeybees can't do it, which is why anyone wanting to grow tomatoes indoors needed to do what the Europeans were doing—develop a steady supply of domesticated bumblebees. Either that, or be prepared to visit every flower in the greenhouse with a buzzing toothbrush.

"For a couple of years in the 1990s, they shipped queens to Belgium for rearing," Robbin explained. Since the Europeans already knew how to raise captive bumblebees for the greenhouse trade, it made sense for American growers to take advantage of their expertise. Single queens, well fed in controlled conditions, could soon

FIGURE 9.1. Greenhouses proliferated during the nineteenth century as the Industrial Revolution brought down the price of glass, firmly establishing hothouse tomatoes as a profitable crop. REPRODUCTIONS © DOVER PUBLICATIONS (LEFT) AND BOSTON PUBLIC LIBRARY (RIGHT).

produce thriving colonies in ready-made cardboard nest boxes that could be shipped anywhere. But when those first Belgian-raised bees came home again, Robbin thinks they brought a European pathogen along with them. "The timing lines up pretty well," he said—a disease outbreak wiped out a bunch of greenhouse bumblebees in 1997, right before the wild species started to disappear. Growers laid the blame for that die-off on a particular and peculiar little organism called a *microsporidian*.

"We've been trying to put the *Nosema* theory to the test," Robbin told me. Then he gave a short chuckle and added, "It shows you how little we know when we can't even decide what Kingdom it's in!" Once considered a protozoan, *Nosema bombi* now finds itself classified as a fungus, or at least something very similar. It's a single-celled organism, shaped like a tiny lima bean, that invades the stomach lining of bees. Infected cells eventually burst, releasing a proliferation of reproductive spores that spread quickly through what amounts to bee diarrhea. Other bees pick up the spores inadvertently at soiled flowers, or from droppings within the nest itself. (In honeybees, young workers charged with tidying up the hive often have the highest rates of infection.) Many bumblebee species seem to tolerate the parasite, at least at low levels, and studies of museum specimens show that it has been widespread across North America for centuries. But for some reason both the rate and intensity of infection seems to have spiked in that one subgenus so closely related to Franklin's. And many of those species are in steep decline. No one can yet say exactly what's going on, but research from the Bee Lab at Utah State University in Logan may explain why the populations have dropped so fast. *Nosema* doesn't just make the bees sick; it prevents them from having sex.

"The queens and workers don't seem too bothered by it," said Jamie Strange, a research entomologist with the Bee Lab. "But the males get so full of spores they can't fly. They just skip across the ground." For ten years, Jamie and his colleagues have been studying western bumblebees in captive colonies, where the plight of the drones can be seen up close. Trouble getting airborne is just the beginning. As the infection worsens, their abdomens become so grossly distended they can no longer curve downward to touch a receptive queen in the appropriate spot. "They can't mate," Jamie summarized. "And when that happens, you're done. In a couple of generations, it all falls apart."

The explanatory power of Jamie's theory is compelling, and it fits the body of evidence nicely. If *Nosema* only weakened bees, it might deplete their numbers over time, but stopping reproduction would wipe out whole populations in just the way Robin Thorp had observed in the field. Here one day, gone the next. Still, the bloated males offered only a mechanism for the declines, and Jamie hadn't yet bothered writing up his observations for publication. All the larger questions remained unanswered.

"Why do some species get sicker than others?" he mused, noting that most bumblebees seemed to shrug off *Nosema*, and even species in the affected subgenus had reacted differently. The Franklin's bumblebee had disappeared, and its rusty-patched cousin was now so scarce that it had recently been added to the US endangered species list. But certain populations of the western and yellow-banded bumblebees appeared to have stabilized, and a fifth member of the group, the white-rumped bumblebee, had apparently never suffered much in the first place. Were some populations naturally more resistant? Was there something different about their environment, or their behavior? And if *Nosema* had always been common, as museum specimens indicate, then why had it suddenly become so deadly? Robbin Thorp still suspected a virulent foreign strain, but genetic tests hadn't yet revealed anything different about the *Nosema* variety found in greenhouse bees. The same pathogen appeared to be having wildly different impacts on different bees in different places.

"It's extremely complicated," Jamie said. "Answers may take a while." Then he made the comparison to human pathology, where scores or even hundreds of well-financed research teams often devote years or decades to unraveling the workings of a single disease. "We can't exactly bring those kinds of resources to bear," he observed, perhaps a little wistfully. But that's not to say that the Bee Lab isn't a busy place. Officially known as the Pollinating Insect Biology, Management, and Systematics Research Unit, the lab gives home to half a dozen full-time bee scientists and three times that number of support staff, as well as a steady stream of graduate students and post docs. I reached Jamie by phone and we talked at length, but the only reason we weren't interrupted is because he'd planned ahead and shut his office door, so that everyone would think he'd gone to lunch.

In addition to the *Nosema* project, Jamie's group was analyzing a broad spectrum of pathogens from four thousand bumblebee specimens gathered across forty sites in sixteen states. Fungal diseases were widespread, but they had also found viruses, mites, bacteria, and a nematode that invaded and destroyed the reproductive organs of queens. There were protozoans and parasitic flies, and the larvae of a beetle that attached itself to bumblebee feet, hitching rides from flower to flower. "When we're finished we'll have a spectacular dataset of what infects bumblebees," Jamie said. Ultimately, they hoped to be able to associate individual pathogens with particular symptoms in particular species—a vital first step in understanding how something like *Nosema* could suddenly become so deadly. Their results would amount to a disease library, helping researchers know what to look for the next time a population of bumblebees started to wane. The work was important, because with all the modern threats to bees and their habitat, most experts think the next decline is probably just a matter of time. After all, even the best-known, best-cared-for, and most widespread bee in the world has found itself in trouble lately. Domestic honeybees have always suffered a few annual losses from what beekeepers traditionally called "Dwindle Disease." But when hives started winking out en masse in the fall of 2006, it was clear that the problem required a new name.

"We were all kind of blown away by what we were seeing," Diana Cox-Foster told me, remembering the first months of the crisis. She now works alongside Jamie Strange at the Bee Lab, but Diana was a professor of entomology at Pennsylvania State University when the die-offs began. "This was not ordinary bee loss," she recalled. Instead of a slow attrition, whole populations of honeybee workers were simply disappearing. They went on their foraging trips—looking apparently healthy—but failed to return, leaving behind combs full of honey and brood, a few disoriented hive bees, and an untended, dying queen. Called into action by distraught beekeepers, Diana devoted her insect pathology lab to analyzing samples from scores of emptied hives. Soon she was collaborating with researchers from New York to Florida, and there were reports of heavy losses on the West Coast as well. At the annual meeting of the American Beekeeping Federation, Diana and her far-flung colleagues gathered one evening at the hotel bar to compare notes. Someone suggested that the word "collapse" fit the situation a lot better than "dwindle," and they all agreed that calling it a "disease" was incorrect, and possibly even misleading. That word implied that hives were falling prey to a particular condition or pathogen, when in fact nobody had any clear idea what was causing the sudden declines. Discussion continued, and by the time they left the bar they had agreed on a phrase that would soon bring the plight of bees to international attention: Colony Collapse Disorder (CCD).

"We wanted a name that would describe the situation accurately, and also create a path forward," Diana explained, and it's safe to say they were successful on both fronts. Press reports about beekeepers losing 35, 50, or even 90 percent of their hives captured the public

FIGURE 9.2. Colony Collapse Disorder can empty an apparently healthy honeybee hive like this one in a matter of days. Without warning, thousands of workers simply fail to return home, leaving behind a few disoriented hive bees and an untended, dying queen. IMAGE BY BOOKSCORPIONS VIA WIKIMEDIA COMMONS.

imagination, earning CCD an even catchier media nickname, "the Beepocalypse." All that attention helped trigger what can only be described as the largest surge of bee research in history. Experts from universities, government agencies, and industry groups quickly launched programs to study CCD, exploring the effects of everything from pathogens (Diana's specialty) to climate change to the signals given off by cell-phone towers. After more than a decade and hundreds of peer-reviewed research papers, the phenomenon is still best described as a disorder—no clear "smoking gun" has emerged as the driving factor. While some of the quirkier notions have been discarded (cell towers and sunspots), many other theories remain subjects of ongoing investigation. The challenge lies in parsing all the possible impacts to a hive, whose members are capable of ranging over an area of 50, 100, or even 200 square miles (130, 260, or 520 square kilometers). Inconclusive—and sometimes downright contradictory—results have sparked heated debate, but there is growing consensus that CCD is caused by a combination of problems. Some people have gone so far as to suggest giving the malady yet another new name: Multiple Stress Disorder.

When I asked Diana what she thought about the Multiple Stress idea, her answer was nuanced. "It does appear to be caused by an interaction of the factors," she agreed, but added that bees weakened by a suite of problems probably succumbed to disease in the end. She cited a greenhouse study where worker bees infected with high virus loads invariably left their artificial hives to die in the far corners of the enclosure. Transfer that habit to the field, and it would look a lot like colony collapse, with ailing bees simply flying off and vanishing into the surrounding countryside. (This aspect of the phenomenon underscores one of the main challenges of studying CCD—it leaves behind scant evidence, like a murder investigation without any bodies.) But then Diana surprised me, going on to point out that over the past few years, documented cases of bona fide CCD have actually become quite rare.

"Less than 5 percent of recent die-offs have the defined symptoms of true Colony Collapse Disorder," she told me. Yet beekeepers across North America continue to lose over 30 percent of their hives every year, and the rate in Europe is also abnormally high. I spoke with several other researchers and all agreed that honeybees are suffering from something much broader than CCD alone. For all its fame, the Beepocalypse appears to be only part of the problem, and one that has left a lot of unanswered questions. What made it spike in 2006, and why is it waning now? What particular stressors cause it, and why are some hives more susceptible than others? And why did it strike widely across North America and Europe, but less so in South America, Asia, and Africa? These and other mysteries of CCD may never be fully understood, but there is a silver lining. The spate of research it stimulated has given scientists a better understanding than ever before of the overall health of bees and the many threats they face in our modern, human-dominated landscapes.

"We talk about the four Ps," Diana told me: "parasites, poor nutrition, pesticides, and pathogens." I had reached her by phone, and she delivered this explanation in the measured tones of someone who is used to talking about her research, but perhaps also a little wary of being misunderstood. With something as complicated and contentious as honeybee declines, it was easy to understand why. Nevertheless, the case she laid out for me was crystal clear, beginning with the story of a nasty little creature that looks something like a red pepper flake, if red pepper flakes came equipped with eight grasping legs and a mouth like a sharp, two-pronged straw.

"*Varroa* is still a major issue," Diana said, referring to the parasitic mite *Varroa destructor*. It lives almost exclusively on honeybees, part of a small group of mites named for the Roman statesman and scholar Marcus Terentius Varro, who, in addition to serving as Julius Caesar's librarian, drafted a theory known as the Honeycomb Conjecture. A beekeeper himself, Varro marveled at the perfect, equal-sided hexagons of his bees' combs. He proposed that they built them that way for the sake of efficiency—that no other interlocking shape could hold so much honey while using so little wax. When a mathematician finally proved that notion right, in 1999, he did Varro a great honor—more so, perhaps, than the mite taxonomist who came up with the genus *Varroa* in the family Varroidae, thus forever associating the old Roman with a deadly threat to the bees he so admired.

Varroa destructor mites make their living by sucking the body fluids from honeybees. They will attack and weaken adults, but they do far more damage inside brood cells, feasting on larvae. Diabolically, that's where they reproduce, proliferating within a sealed chamber right alongside their defenseless prey. In Varro's time, they existed only in the forests and woodlands of Southeast Asia, where they were minor pests of various native honeybee varieties. (Of the eleven currently recognized species in the genus *Apis*, only the domesticated *Apis mellifera* is native to Africa and Europe. The others are Asian.) But they quickly adapted when domestic honeybees arrived in the region, and then spread steadily around the globe with the movement of hives, queens, and equipment. Now they are

FIGURE 9.3. A scanning electron microscope image of a female *Varroa* mite perched on the shoulder of a female honeybee. IMAGES COURTESY OF ELECTRON AND CONFOCAL MICROSCOPY LABORATORY, AGRICULTURAL RESEARCH SERVICE, US DEPARTMENT OF AGRICULTURE.

a major problem everywhere outside Australia. Left untreated, mite infestations can cripple brood production and decimate entire hives, but they also vector several deadly viruses, further weakening the bees wherever they occur. Experts have linked the arrival of mites to the decline of feral honeybee colonies in parts of Europe and North America, and if Diana is right, they weaken the overall health of bees, much like the second P in her model, poor nutrition.

"There just aren't enough floral resources," Diana told me, explaining how the idea of poor nutrition had made it onto her list of four Ps. "People look across a park or a golf course and think it's green and lush, but to a bee it's like a desert or a petrified forest—there's nothing to survive on." In addition to the scarcity of flowers in parks and the loss of natural areas to development, habitat is also eroding in agricultural settings, where hedgerows and the mixed crops and pastures of traditional farms have been increasingly replaced by monocultures. Even nectar and pollen-rich weeds like thistles and broom and bindweed have gotten harder to come by in many places, controlled now with herbicides everywhere from farms to backyards to roadsides.

Diana's comments echoed something I heard from Dr. Larry Brewer, a contract researcher whose firm maintains hundreds of beehives and conducts large-scale field trials for agrochemical companies. Testing the effects of new products often involves isolating those hives out in the middle of huge fields of canola or other bee-pollinated crops. But even at the peak of the bloom, Larry's team always finds at least some bees returning to the hive with other kinds of pollen. "They'll fly a long way to find what they need to find," he said, pointing out that bees seem to crave things they can't get from only one kind of flower, no matter how prolific it might be. "They go looking for other sources of protein and micronutrients, even when they're surrounded by what looks like a perfect meal." The nutrition problem is particularly challenging for commercial hives that get trucked from one monoculture to the next over the course of a season. "Just imagine if that was your diet," Larry said— weeks of almonds, followed by weeks of apples, followed by weeks of nothing but blueberries, with every stage punctuated by a long road trip confined to the hive. Beekeepers provide supplements, but nothing can completely replace what honeybees evolved to eat— the varied floral rewards from a diversity of wildflowers, shrubs, and trees. While stress from poor nutrition varies considerably over the course of a season, and some hives suffer more than others, experts like Diana believe that it undermines overall health and stamina, making bees more vulnerable to other threats in their environment, including the third and most controversial of the Ps.

No single aspect of bee decline has sparked more debate than the impact of pesticides. But before exploring that issue, it's worth addressing a fundamental question at its root: What makes bees so susceptible to chemicals in the first place? Why don't they ever de- velop resistance to pesticides the way all the target insects seem to? The answer to that puzzle is an intriguing consequence of the special relationship between bees and flowers. For locusts, hornworms, bee- tles, aphids, Lygus bugs, and all the other pestiferous creatures that attack leaves, stems, seeds, and roots, their whole existence relies on detoxifying complex compounds. They've been doing it for millions of years, struggling to overcome the constantly evolving chemical defenses of the plants they feed on. (It's an arms race that pesticide makers are well aware of, and they often look to plants for inspira- tion, tweaking various extracts one way or another to create new products.) But bees are different. Their role as pollinators creates a need for plants to attract them, not repel them, which has led to the evolution of sweet nectar and protein-rich pollen that rarely con- tain any defensive chemicals at all. While this keeps bees well fed, it means they have little evolutionary experience with harmful com- pounds in their diet. They lack the ingrained metabolic pathways that pest insects use to process plant chemicals and find ways around them. For the crop eaters, pesticides amount to a familiar—and usually temporary—chemical setback. For bees they're just a poison, no matter what form they take.

FIGURE 9.4. The chemical war on crop pests—often using plant-based toxins—is a struggle as old as agriculture, perfectly captured here in a poster issued by the US Department of Agriculture during World War II. WIKIMEDIA COMMONS.

"We can't link bee declines to one chemical, or even one class of chemicals," Diana said immediately, as if anticipating my line of questioning. I wanted her opinion on a group of pesticides called *neonicotinoids*, which include some of the most popular agricultural and home garden products on the market. "Neonics," as they've come to be known, can be applied in a variety of forms, but they all share the trait of becoming systemic—they're taken up into the very tissues of the growing plant. This means that the plant's leaves, buds, and roots become deadly to chewing pests, reducing the need for indiscriminate spraying. But it also means that neonics show up in the plant's nectar and pollen, putting them directly into the diet of visiting bees. Nobody doubts that neonics are toxic at high doses—they're designed to kill insects, after all—and cases of bungled applications have caused indisputable local die-offs of honeybees and native bees alike. Laboratory studies have also linked neonics to a variety of what researchers call "sub-lethal" effects, from impaired foraging and homing abilities to shortened life spans and low fertility. But that's where the consensus ends, because no consistent colony-level impacts have been shown on honeybees in the field. Hives raised amid treated crops often appear to do just fine, and proponents argue that the vast majority of domestic honeybees encounter only trace amounts of the pesticides under normal conditions. There is much stronger evidence that neonics harm wild bumblebees and solitary bees, however, and they've also been implicated in declines of other non-target species, including insect-eating birds. With controversy mounting, the European Commission banned several types of neonics from use on flowering crops in 2013, and it's reportedly considering a much wider prohibition.

Like most of the other scientists I talked to, Diana didn't endorse an outright prohibition on neonics. "The push now should be more for integrated pest and pollinator management," she said. "Not necessarily eliminating pesticides, but asking, 'What do I absolutely have to use, and how do I do it to keep bees healthy?'" (When she said that, I thought immediately of how the alfalfa farmers in Touchet were constantly adapting their pesticide management—looking for more bee-friendly products, experimenting with dosages, and applying things only after dark, when the alkali bees were safely in their beds. "We think about it all the time," Mark Wagoner told me.) Nonetheless, farmland affected by the European neonics ban will be an important test case, allowing researchers to gauge not just how bee populations respond, but also the impacts of whatever chemicals are brought in as alternatives. In the meanwhile, Diana and others have learned that neonics make up only one facet of a much more complicated pesticide picture.

"We were amazed," Diana said, recalling the results of the first large-scale analysis of chemical residues in pollen, honey, wax, and the bodies of bees. Samples from scores of hives across North America yielded 118 different pesticides—not just modern varieties, such as neonics, but things that had been lingering in the environment for years, or even decades. "Basically, everything that's ever been used," she told me, and for the first time in our conversation, her voice betrayed something like outrage. "DDT is still coming in on pollen!" The contaminants included fungicides, herbicides, miticides, and a wide range of insecticides. But the chemicals weren't just diverse, they were nearly ubiquitous. Among 750 samples analyzed, only 1 piece of wax, 3 bits of pollen, and 12 adult bees tested clean. The rest contained an average of between 6 and 8 pesticides each. And that's where things get really interesting.

"They synergize," Diana told me, "which tends to make them much worse for bees." She explained how cocktails of chemicals could work in combination, one enhancing the effects of the other. Fungicides, for example, don't always harm bees on their own, but they can make certain insecticides up to 1,100 times more potent. Yet products are only tested and evaluated by regulators one at a time, so things labeled "bee safe" for use on their own can have unforeseen consequences in the presence of other pesticides. With bees encountering so many chemicals in so many potential combinations—the vast majority of them unstudied—it's no wonder field trials have produced confusing results. Even the so-called inactive ingredients in a mixture can play a role. When we spoke, Diana and her colleagues had just determined that a common surfactant used to enhance liquid neonic applications had an unexpected side effect: it doubled the death rate of bees infected with viruses. So agrochemicals don't just interact with each other, they can also compound the effects of pathogens—the final, and in some ways most threatening, of the four Ps.

"The honeybee is basically the poster child for disease in insects," Diana said. "Everything you can see in humans—from viruses to bacteria to protozoans—is also found in bees." She rattled off a list of pathogens with descriptive names like deformed wing virus, acute paralysis virus, and chalkbrood. There was a honeybee form of *Nosema*, and a dreadful-sounding bacterial condition called foulbrood that basically reduced combs full of bee larvae to a stinking, black goop. Talking about *Nosema* again reminded me of the bumblebee situation, but with honeybees there is now less guesswork. The spate of research inspired by CCD resembles the kind of massive epidemiologic effort that Jamie Strange dreamed about—for viruses alone, over twenty new honeybee varieties have been isolated and named. But longtime observers like Diana still don't understand why the situation seems to be getting worse. "As recently as 2000, you could find hives with no trace of viruses," she recalled. "Now they all have them." There is also evidence that honeybee pathogens can jump to bumblebees or other native species, a particularly troubling development with so many hives and queens still being trucked and shipped around the world. Just like the *Varroa* mite spreading from its home in Southeast Asia, so, too, have many honeybee diseases started off as local problems, a trend revealed by distinctly geographical names like Kashmir Bee Virus and the Lake Sinai Virus. But most experts believe the knowledge amassed from honeybee research will eventually help all bees, a hope that influenced Diana's decision to leave her university post and join Jamie Strange and other native bee researchers at the Bee Lab in Utah.

"Native bees present a challenge," Diana admitted, describing her shift to working with species like bumblebees, mason bees, and alkali bees. She noted how much more difficult they were to rear in a lab, and how their life cycles were short and seasonal, preventing the kind of year-round research possible with honeybees. "But we have enough preliminary data to say that the four Ps apply to them as well." Given the chance, some experts might add a few more letters to that model—an N for the nesting habitat so easily lost to development and industrial farming; an I for invasive species (both bees and plants); and a double C for an overarching issue with the potential to complicate everything—climate change. Bee experts are just beginning to explore the impacts of climate change, but there is an obvious risk that earlier spring blooms will leave some bees in the lurch, emerging from their nests and hibernacula too late to find their preferred sources of nectar and pollen. No one yet knows how quickly bees will be able to adjust, but a study of North American and European bumblebees found them retreating from hotter temperatures in the southern and low-lying parts of their ranges, but failing to take advantage of milder conditions in the warming north. Extreme weather events are also on the rise: as Mark Wagoner explained to me on his alfalfa farm, it just takes one big thunderstorm at the wrong moment to drown an entire population of alkali bees. Other species may be equally vulnerable to the frequent droughts, flooding, heat waves, forest fires, and unseasonable cold snaps expected in the decades ahead.

Taken together, the four Ps (plus N, I, and double C) paint a challenging picture for bees in the twenty-first century. Some species, like the Franklin's bumblebee, may already be extinct, and many more have disappeared from at least parts of their range. To say that all bees are in decline would be going too far, but there is now a cautionary tale about what things might be like if they were. Beginning in the 1990s, the renowned apple orchards in China's Maoxian Valley saw a drop in bee populations that quickly turned into a complete collapse. Nobody knows exactly what happened, but most observers blame an excessive and careless use of pesticides, combined with the poor nutrition and lack of nesting sites associated with habitat loss. Wild bees effectively vanished, and domestic colonies failed again and again, until beekeepers simply refused to bring more hives into the valley. To save themselves from ruin, local orchardists began hiring thousands of seasonal workers to hand-pollinate their trees. It was a painstaking process. Unlike date palms, where hundreds of tiny flowers can be swabbed with a single cotton ball, apple trees require an individual dab of pollen for every blossom. Armed with a long stick topped by chicken feathers or a cigarette filter, even the fastest workers could service only five or ten trees in a day. Not surprisingly, the practice proved unsustainable economically—human labor simply couldn't replicate what the bees had always done for free. Farmers began cutting down apple trees en masse and replacing them with other crops. Today, all that remains of a once-thriving apple industry are a few orchards at the very edges of the valley, where bees surviving in nearby forests can help pollinate the trees.

In an interesting twist, many of the Maoxian Valley farms once devoted exclusively to apples have now adopted a more traditional, mixed-cropping approach, growing loquats, plums, and walnuts interspersed with vegetables. Though the change was driven largely

FIGURE 9.5. In China's Maoxian County, habitat loss and pesticides led to a collapse of local bee populations. Orchardists responded by hiring teams of human pollinators to service every flower, carefully applying pollen with long sticks topped with feathers, or, in this picture, cigarette filters. PHOTO © UMA PARTAP.

by economics, it may also help bring back the valley's bees, because the landscape now offers greater diversity of pollen and nectar in a setting less dependent on pesticides. So while the Maoxian story is often held up as a warning about bee declines, it may ultimately become a symbol of bee resilience, a reminder that solutions are within our grasp. Among the many experts I contacted while researching this chapter, I heard the most practical advice from bumblebee specialist Dave Goulson, a professor of life sciences at the University of Sussex. He agrees that multiple stressors affect bees in complex ways, but argues that we don't need a full understanding of the problem to do something about it. "This is not an excuse for inaction while further research is performed," he wrote to me in an email. "Common sense suggests that reducing pressure from any of these stressors will help." In short, we know enough to act, and we know enough to act in specific ways: by providing landscapes with more flowers and nesting habitat, reducing pesticide use, and stopping the long-distance movement of domestic bees (and the pathogens that travel with them). Putting even some of these straightforward ideas into practice can be transformational, as a growing number of scientists, farmers, gardeners, conservationists, and ordinary citizens are quickly beginning to learn.

CHAPTER TEN

A Day in the Sun

In this flowery wilderness the bees rove and revel, rejoicing in the bounty of the sun, clambering eagerly through bramble and hucklebloom, ringing the myriad bells of the manzanita, now humming aloft among polleny willows and firs, now down on the ashy ground among gilias and buttercups, and anon plunging deep into snowy banks of cherry and buckthorn. They consider the lilies and roll into them, and, like lilies, they toil not, for they are impelled by sun-power, as water-wheels by water power; and when the one has plenty of high-pressure water, the other plenty of sunshine, they hum and quiver alike.

—John Muir,
"The Bee-Pastures of California" (1894)

I didn't expect the vacuum cleaners. When I flew to California to visit an almond orchard, I knew I would see nut production on a grand scale. With over 940,000 acres (380,000 hectares) devoted exclusively to almond trees, California's Central Valley produces a whopping 81 percent of the world's annual harvest. Every summer, hydraulic shakers move through the orchards, grasping trunk after trunk with their padded arms and vibrating ripe nuts to the ground in a flurry of dust and leaves and desiccated husks. If that's all there were to harvesting almonds, then the orchards might have been full of native bees. They could have feasted on the almond bloom as early as February, and then enjoyed a mixture of wildflowers and cover crops in the understory as spring progressed into summer. But as we drove north from Sacramento and began encountering orchards alongside the highway, I could see immediately why bee conservation was an issue for the almond industry. There was nothing whatsoever growing beneath the trees—not a flower, not a weed, not a blade of grass. Heavy mowing and herbicides hadn't simply reduced the vegetation, they had removed it entirely, leaving behind a powdery brown moonscape of bare soil.

"It's because of the harvest—they need to vacuum up the nuts," said my guide for the day, a pollinator specialist named Eric Lee-Mäder. He explained how a mechanized sweeper followed the tree-shakers, arranging the scattered nuts into neat windrows that could be hoovered up by yet another machine trundling along behind. While highly efficient, this process required the cleanest ground possible, and it's not surprising that industry manuals refer to the space below an almond tree as the "floor." Allowing plants to clutter things up would just make the nuts harder to gather—like trying to pluck crumbs from a thick shag carpet. What's more, vegetation would provide cover for nut-eating rodents and trap water that increased the risk of *Salmonella* and other contaminants infecting the almonds. Maintaining tidy floors helped growers keep the harvest sterile and streamlined, but it also meant that California's vast almond orchards contained almost no habitat for bees. And that had consequences for a crop that absolutely required bees for pollination.

"We're working in over 10,000 acres [4,000 hectares] of almonds now," Eric told me, noting a surge in the number of growers who

FIGURE 10.1. The immaculate floor of a typical almond orchard may be convenient at harvest time, but it leaves little habitat for bees. IMAGE COURTESY OF USDA NATURAL RESOURCES CONSERVATION SERVICE VIA WIKIMEDIA COMMONS.

wanted to do something positive for bees. As codirector of pollinator conservation for the Xerces Society, he was in a unique position to help them. Founded in 1971 and named for an extinct California butterfly, Xerces is the only major nonprofit in North America devoted to saving insects and other invertebrates. Eric joined the staff in 2008, just as Colony Collapse Disorder was bringing the plight of honeybees to international attention. The group has been expanding ever since, fueled largely by growing public concern for pollinators. "I think I was the fifth or sixth hire," he said. "Now there are fifty of us." That same trend has helped two similar organizations take root and thrive in the United Kingdom: Buglife (founded in 2002), and the Bumblebee Conservation Trust (founded in 2006). Collectively, these groups have helped transform the growing awareness of bees into specific actions—for example, adding Hawaii's masked bees to the US endangered species list, improving pesticide policies, and establishing the world's first bumblebee sanctuary at Loch Leven, Scotland. For years, I had been following such developments with the sort of detached enthusiasm one gets from writing an annual donation check. But now I found myself wanting to know more. What exactly did it mean to "promote habitat conservation and restoration" for bees, and, more importantly, did it work? When Eric invited me to join him for a day in the field, I jumped at the chance.

"Today we'll see the beginning and the endpoint," he said as we drove, passing more and more almonds, as well as pistachios, olives, and occasional fields of sunflowers, tomatoes, or rice. We were running a little late for our first stop, an orchard just getting started with bee conservation, and then Eric planned to show me a well-established, mile-long hedgerow, one of the first projects he'd worked on for Xerces. Near the small town of Orland, we turned off the highway and succeeded in baffling our GPS navigation system on a series of look-alike roads through the orchards. "There's a native plant!" Eric said suddenly, and put on the brakes. The sight of a ditch full of blooming gumweed told him we'd found our destination.

Two of Eric's colleagues had already arrived, and we caught up with them on a dusty berm that separated the road from the orchard's ordered grid of trees. They were deep in conversation with a tall, broad-shouldered man who, in other surroundings, might easily have been mistaken for a professional athlete. He was Bradley Baugher, a fourth-generation farmer whose family owned this large grove and many others, controlling a sizable share of the booming market for organic almonds. One of their largest customers, General Mills, had recently mandated that its suppliers incorporate pollinator conservation into their production stream, and the Baughers had reached out to Xerces. "They were really receptive to the idea," Eric had told me, and indeed, Bradley had been experimenting with native plants on his own for years. The gumweeds in the ditch were his, and he'd also seeded the berm with a mixture of lupines, poppies, *Phacelia*, and *Clarkia*. It was midsummer now, and most of those early bloomers were dried husks, but some of the poppies and morning glories still looked fresh. As everyone shook hands I noticed an auspicious sign: the dusky, spotted wings of a common buckeye butterfly moving among the bright petals.

"There are three features we've used in almonds and had some success with," Eric began, and told Bradley how a combination of hedgerows, native cover crops, and strip plantings could help bring bees back to Baugher Ranch. Eric spoke with a combination of warmth and confidence that seemed to put everyone at ease. He was in his mid-forties, with a direct gaze, cropped hair, and a certain professional polish that may have been a holdover from his previous career in the tech industry. "I'm the token capitalist at Xerces," he joked later. But while his colleagues may have had the entomology degrees, Eric's bona fides ran deeper. He'd grown up on a North Dakota farm in a family of beekeepers. Though he'd done many other things in the years since, those roots were a bridge to the people he was working with now. "I try to build real friendships," he confided. "Establishing trust is the biggest challenge in this work."

For his part, Bradley Baugher appeared to approach the day with cautious optimism. He showed a genuine interest in establishing more bee habitat, and, like any farmer, a curiosity about which plants would get the job done. The conversation moved from various wildflower mixes to blossoming shrubs and weed control as we toured the field edges, abandoned ponds, and other unused corners of the ranch available for planting. But he also had a farmer's concern for practicalities and the bottom line, bringing the discussion back to earth with comments like, "We need to avoid anything that would compete with the almond bloom," or, "My crew would need two full days to weed-whack all of that." Lunchtime found us in the pleasant air-conditioned staff room at ranch headquarters, a welcome relief from the 95-degree heat (35 degrees Celsius) outside. But Bradley called the weather cool and said it would be a lot worse in a few weeks. "113 degrees—the perfect time to harvest!" he laughed, quoting an old family saying. Over fresh watermelon picked from his garden that morning, we learned more family lore—how his great-grandparents had brought their mules across the country by train, and how he and his wife would soon be adding a baby to the fifth generation of Baugher farmers. Finally, talk turned back to bees, and one of the main challenges in growing almonds: getting all those thousands of trees pollinated.

"Bee availability has been a struggle since the die-off," Bradley admitted. He had the sort of open, honest face that would have been terrible in a poker game, and it was clouded now with worry. But any other almond grower might have looked the same way—pollination was a gamble every year. With so few resident bees to speak of, California's orchards have long depended on rented pollinators to ensure a viable crop. Commercial beekeepers come from as far away as Florida and Maine to participate in the world's most intense and lucrative pollination marketplace, a three-week frenzy of honeybees and almond blooms. At the recommended stocking rate of two units per acre, California growers require more than 1.8 million hives to service their trees. But meeting that demand has become increasingly difficult—bee supplies have simply never recovered since the outbreak of CCD. Hives that rented for fifty dollars a decade ago can bring four times as much today, making them so valuable they've become the target of what news stories refer to as "bee rustlers." Thousands of hives now disappear from orchards every year, spirited away in the dead of night to be repainted, rebranded, and pawned off to a different grower. The amount of money involved can be surprising. In 2017, police arrested two men tending a cache of contraband bees worth close to a million dollars.

With the stakes so high, it's no wonder that more and more almond farmers are exploring the potential of native bees. Eric is quick to point out that planting a few flowers and hedgerows isn't a panacea—even the most bee-friendly orchards still rent honeybees every year. But studies have linked the presence of wild species to an increase in fruit set, and shown that planting natural vegetation can quickly triple the diversity of pollinators in an orchard. The additional flowers benefit honeybees, too, improving nutrition and reducing the stress of constant movement. Beekeepers appreciate (and seek out) orchards where their hives can remain in place after the almond bloom, feeding on a wider variety of pollen and nectar. In the parlance of conservation, bee habitat pays "stacked environmental benefits"—supporting a range of beneficial insects and other species while also sequestering carbon, increasing soil moisture, and adding to the soil's organic matter. But the decision to participate often boils down to something more basic and intangible: in Bradley's words, helping bees was "the right thing to do," and Baugher Ranch wanted to set an example. He and Eric spent a lot of time discussing how to make plantings visible and attractive from the main road. "We want people to see this," Bradley said.

By the time we left Baugher Ranch, we'd met Bradley's mother and sister, talked about tractors with his brother-in-law, and eaten fresh peaches from a tree behind the family home. A clear plan for bee habitat had also emerged: Xerces would provide technical expertise and help defray the cost of seeds, while Baugher Ranch provided the labor. Planting roadside strips would receive first priority, followed by fence-line hedgerows and several acres of old ponds and pasture. With those projects underway, Eric thought the ranch would be a shoo-in for a new certification program called "Bee Better." Modeled after the organic and fair trade movements, it's designed to add a recognizable label (and value) to bee-friendly products. Bradley promised to look into it as we said goodbye and piled back into our rental car. "I'm glad to have you guys on board," he said in parting, and I didn't remind him that I was just an observer. It felt good to be mistaken as one of the team.

Eric and I both had planes to catch, but there was just enough time left to stop by the mature hedgerow he wanted to show me. "In this landscape, the change is jaw-dropping," he explained. "To see what was quite literally dust become flowers, bursting with life. . . . It's astounding." In addition to the expected bees, Eric had spotted all sorts of other creatures visiting his restoration sites, from hummingbirds and butterflies to coyotes, pheasants, snakes, and raptors. Once, a peregrine falcon had snatched a starling from midair directly above his head. "I can't fathom where these animals come from," he said, and as we drove along through mile after mile of orchards and fields, farmed right up to the edge of the road, I could understand his surprise. Hardly a scrap of natural vegetation remained in what naturalist John Muir once called the world's greatest bee-pasture. Recalling his first visit in the spring of 1868, Muir described the valley as "one smooth, continuous bed of honey-bloom, so marvelously rich that, in walking from one end of it to the other, a distance of more than four hundred miles, your feet would press more than a hundred flowers at every step." After a century of intense cultivation, the fact that native bees and other wildlife remained at all was profoundly hopeful, as if an echo of Muir's wild pasture lay just out of sight, ready to flourish wherever an oasis of flowers could be established.

When we finally arrived at the hedgerow, Eric's tone suddenly turned apologetic, as if, after all the buildup, he feared that I might be disappointed. It was too late in the season to see many bees, he warned, and this particular hedgerow had faced its share of problems. "It's actually sort of a plagued site," he said, and rattled off a list of setbacks, from flooding to errant road graders to a drunk driver who plowed through a large section of young plantings. But in spite of all that, I could see that the hedgerow was working before I even got out of the car. It stretched in a lush line along the road edge, like a green wave breaking on a desert beach. Ceanothus, wormwood, and saltbush grew head high, interspersed with clumps of perennials like wild buckwheat and yarrow. Their leafy shade stood in sharp, almost comical, contrast to the other side of the road—a dusty, bone-dry verge dotted here and there with the skeletons of star thistle. We pulled over, and while Eric took a call on his phone, I stepped out into the heat for a closer look.

In July, even John Muir would have struggled to find flowers in a Central Valley hedgerow. Scorching, droughty weather made summer what he called the "season of rest and sleep" for local plants, and I wasn't surprised to see that most of the shrubs and perennials had long since gone to seed. But their greenery still pulsed with life. I saw spiders and wasps, and a profusion of slender dragonflies perched on the twigs and branch tips. A kingbird called squeakily overhead, and I heard a mockingbird riffing on its song from somewhere behind an elderberry bush. Then I spotted a patch of gumweed still in bloom, the same species Eric had noticed in the ditch at Baugher Ranch. Its yellow flowers glowed in the sunlight, and within minutes, two checkered skippers and a cabbage white butterfly had stopped in for nectar. And then came a bee, a small, shiny sweat bee with a neatly pinstriped black and white abdomen. The pollen clinging to her back legs told me she was still provisioning nest cells somewhere nearby, and I watched while she busily scraped and prodded, adding more golden grains to her supply. In another setting, this sight would have been unremarkable—a native bee acting normally on a native flower. But here, isolated in one of the world's most intensively farmed landscapes, that little bee struck me as a powerful symbol of resilience, and of the potential to restore bees pretty much anywhere. It's not surprising that Xerces partners with all kinds of landowners, creating new bee habitat everywhere

FIGURE 10.2. A native sweat bee foraging on a native wildflower, hopeful signs of bee restoration in California's Central Valley, one of the world's most intensively farmed ecosystems. PHOTO © THOR HANSON.

from backyards and gardens to golf courses, parks, and airports. "Anyone can do this," Eric said at one point, a sentiment I heard again when I spoke with his boss at Xerces, executive director Scott Hoffman Black.

"I've been doing conservation work for a long time," Scott told me over the phone. "I've worked on wolves, salmon, spotted owls. . . . But this is the first time I've been able to show people how to get results they can see immediately." That sense of instant gratification is in some ways a product of scale. Because bees are small and reproduce rapidly, they can respond quickly to small changes. Many species need only a safe nesting site and a few weeks' worth of flowers to thrive. But while that can make the work of bee conservation satisfying, it doesn't reduce the scale of the challenge. Hedgerows and other habitat projects may be catching on, but Eric and I still had to drive for over an hour through farm country to see one. And then there are all the other problems that groups like Xerces are working on, from pesticides and disease to climate change. When I asked Scott if he was hopeful for the future of bees, he gave a wary chuckle and quipped, "It depends on the day."

I put the same question to Eric Lee-Mäder as we continued on to the airport. There was a long pause, and then he answered indirectly, talking again about the inspiration of connecting with people. He said he found hope in seeing farmers and other landowners embrace conservation and become champions for it. One orchard he worked with had turned into an unlikely showpiece—it was now surrounded and crisscrossed by six miles of hedgerows and habitat features. Unlike Baugher Ranch, however, it wasn't an organic, family-run operation—it was owned by an international agribusiness conglomerate based in Singapore. "They were reluctant at first," Eric admitted, but when those initial plantings started to blossom and buzz, attitudes quickly shifted from skepticism to enthusiasm, and the program has been growing ever since. It will take more than a few hedgerows to secure a future for bees—Eric's colleagues at Xerces also run programs aimed at reducing pesticide use, protecting wild habitat, and saving endangered species. But while the process might include long-term policy efforts and abstractions like "stacked benefits," it can also be as tangible and rewarding as watching a bee on a flower, and the best hope may lie in helping more and more people make that discovery. "I like to think I'm handing them a beautiful painting for their wall," Eric said, in an apt summary, "one that they didn't even know existed."

CONCLUSION

The Bee-Loud Glade

And now I wander in the woods,
When summer gluts the golden bees…

— William Butler Yeats,
"The Madness of King Goll" (1889)

O n the island where I live, people gather for several days every August to partake in a classic celebration of rural living, the county fair. All the traditional activities take place, from carnival rides and a livestock auction to a huge range of competitive events. Horsemanship always draws a crowd, but people also pack the stands for the chicken race, the pie-eating contest, and a fashion show of clothing made entirely from trash and recyclables. It's possible to win a ribbon (and a little prize money) for everything from scarecrows to flower arrangements, and over the years my family has done pretty well on watermelons, beans, red currants, and canned salmon. Every fair revolves around a particular theme, and this year the organizers chose to focus on bees. Posters now appearing around town feature five bright bumblebees flying over a backdrop of sunflower and clover, with a big dripping dollop of honey and the new fair slogan, "It's all the Buzz!"

In a small community people tend to know each other's business, so I wasn't surprised when someone from the fair called to ask if I would give an afternoon lecture on bees. I agreed, but suggested that instead of a presentation, I take people out into the fairgrounds and introduce them to all the various bees that lived there. This proposal earned the kind of pause that, in a phone conversation, amounts to a blank stare. But eventually it was agreed, and a few days later I stopped by for the kind of advance scouting trip that a bank robber might call "casing the joint."

Less than two weeks remained until opening day, and the fairgrounds were bustling with activity. I saw painting crews touching up the various barns and outbuildings; the poultry and rabbit tents were up; and a local sculptor had erected a two-story metal beehive overlooking the horse arena. At first glance, however, I could understand why my bee-walk idea had met with skepticism. Anything not covered by structures was either a parking lot or a field of close-cropped grass, withered and browned by the sun. But then I noticed the scattered yellow blossoms of a weed called cat's ear, and it didn't take long to find sweat bees on them, and some kind of tiny black miner. In a bit of landscaping by the fair's main office, I spotted honeybees foraging on an ornamental sumac. And when I rounded a corner between the bathrooms and the food court, I stumbled onto a bed of lavender where three species of bumblebees and a feisty little wool-carder were madly working the fragrant, purple blooms. I have no doubt they will still be at it when the fairgoers arrive, darting unnoticed through the crowds with single-minded purpose, overlooked players in a vital drama that surrounds and sustains us.

Learning about bees may feel like something new, but the journey has less to do with discovery than rediscovery. People have always lived near them and among them—only recently did we stop paying attention. Allowing bees back into our awareness rekindles old connections, and the results can be profound. I have a friend whose wife died young and unexpectedly, succumbing to a rare cancer mere weeks after feeling the first symptoms. She had been a beekeeper, and when he and his daughter returned home from the hospital, they found her hives in turmoil. The workers were busy tending new queen cells, and within a few days they swarmed, rising in their tens of thousands and massing on a maple branch twenty feet from the front door. He watched that swarm for hours, and later wrote about the experience movingly, calling it "magical, spiritual, otherworldly and comforting." Though remarkable, this episode would once have seemed ordinary, even expected. Throughout Europe and North America, people used to follow a practice called "telling the bees," keeping their hives up to date on all manner of news, from the state of their crops to births, weddings, or illness in the family. Whenever someone died, the bees would be soothed with singing, their hives draped in mourning black. Failure to do so risked offense, and everyone knew that resentful bees would soon swarm and depart. In that era, not so long ago, it was also common to take comfort in the presence of bees, a solace William Butler Yeats famously pined for in "The Lake Isle of Innisfree":

> Nine bean rows will I have there, a hive for the honey bee,
> And live alone in the bee-loud glade.
> And I shall have some peace there . . .

Wherever we find them, bees hum and bustle with the vitality of life itself, and while we may enjoy their honey and appreciate their role in pollination, our affinity for them is something more than practical.

Rachel Carson's Silent Spring gave the environmental movement its most powerful metaphor, a world without birdsong. But she also warned of blossoms without the drone of bees, and there are land-scapes where that vision is already too close to the truth. Much de-pends on us—taking notice, taking heed, and taking action. In my family, the first bees of springtime remain a highly anticipated event, and not long ago my son and I stood together watching several fresh bumblebee queens warm themselves on a sunny, south-facing wall. Three were colored yellow and orange, and the fourth was deepest black, like an animate ink-drop banded with gold. "Bees are special, Papa," Noah said, and I told him that I agreed. Then he added an observation, with that offhand wisdom of youth, that I knew would have to be the last line of this book: "The world could do without us, but it couldn't do without bees."

Appendix A
Bee Families of the World

With over 20,000 species buzzing around on every continent outside Antarctica, bees are one of the most successful groups of insects in nature. The following pages give just a hint of their diversity, describing the seven bee families recognized in Charles Michener's taxonomic opus, *The Bees of the World*. While some groups are rare, many of these bees can be seen in backyards, parks, natural areas, farms, fields, and roadsides anywhere.

STENOTRITIDAE (no common name)

This small but distinctive family of bees lives only in Australia and includes approximately twenty species in just two recognized genera. They are all robust and fast-flying, ranging in color from bright yellow to black to metallic green. The biology of this group remains poorly known, but one observer described remarkable mating flights in several species of *Ctenocolletes*, where females continued normal foraging behavior, accumulating full loads of pollen while *still mounted* by the males! Stenotritid bees visit the flowers of many characteristic Australian plants, particularly members of the myrtle family like *Eucalyptus* and featherflowers (*Verticordia* spp.). They are solitary and nest in the ground, sometimes in loose aggregations. The bee pictured below is *Ctenocolletes smararagdinus*, foraging on a *Eucalyptus* blossom.

ILLUSTRATION © CHRIS SHIELDS.

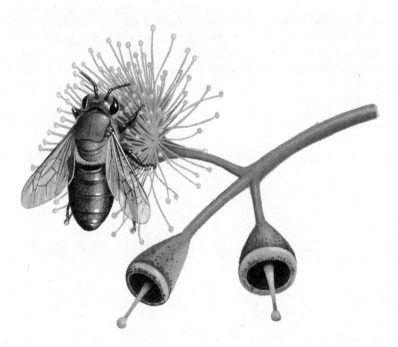

COLLETIDAE—Plasterer Bees and Masked Bees

Widespread and diverse, the 2,000 species in this family include over half the bees found in Australia and nearly nine out of ten species native to New Zealand. Worldwide, the largest and best-known groups are the plasterers (*Colletes*) and the masked bees (*Hylaeus*). Plasterers are fuzzy bees with heart-shaped faces who use their peculiar, bilobed tongues to paint the walls of their nest cells with a waterproof antifungal secretion. When this "plaster" hardens, it forms a clear, flexible lining that looks like something synthetic, earning them the nickname "polyester bees." Masked bees resemble tiny, smooth wasps with patterned faces. They don't need fuzzy legs and bodies because they've developed the odd habit of swallowing pollen and transporting it inside their stomachs. Back at the nest, they regurgitate a mixture of pollen and nectar into each cell and float a single egg atop the slurry. Masked bees are also particularly well-traveled: they're the only variety to have reached the remote Hawaiian Islands, where one ancestral colonizer evolved into at least sixty-three distinct species found nowhere else. Seven of those rare endemics recently became the first bees to appear on the US endangered species list. The bee pictured here is a Eurasian plasterer, *Colletes daviesanus*, with a cutaway of its liquid-filled nest chamber. ILLUSTRATION © CHRIS SHIELDS.

ANDRENIDAE—Mining Bees

Though scarce in Southeast Asia and absent from Australia, mining bees occur nearly everywhere else and number close to 3,000 species. They are particularly common in arid habitats with lots of open ground for digging their nest tunnels, which can reach a depth of nearly ten feet (three meters) for the larger species. All bees in this family are solitary, though some do nest in aggregations and occasionally share the same tunnels. Members of the most diverse genus (*Andrena*—about 1,300 species) boast prominent fringed scopae that run the entire length of both back legs. They often specialize on one or a very few species of flowers, as do the tiny bees in the genus *Perdita* (about 700 species). *Perdita* bees are extremely gentle, and many have lost the ability to sting. Studies have compared the reproductive strategies of mining bees in deserts to the seeds of the plants they rely on. Like seeds, dormant bees can endure in the soil for extended periods, waiting for up to three years for the rainfall that will bring a life-sustaining bloom. The bee pictured here is a female tawny mining bee (*Andrena fulva*) carrying a full pollen load into her nest tunnel. ILLUSTRATION © CHRIS SHIELDS.

HALICTIDAE—Sweat Bees and Alkali Bees

A truly cosmopolitan family, the halictids include over 4,300 species and occur virtually everywhere that bees are found. In hot climates, many varieties are attracted to human perspiration, a habit that gave rise to the common name "sweat bees." The family exhibits a wide range of social behaviors, from entirely solitary species to bees with shared nests, overlapping generations, and a distinct caste of workers. While many sweat bees are small and inconspicuous, others announce themselves with vibrant iridescence. Members of the New World genus *Agapostemon* look like bright green jewels, and the family also includes the alkali bees (*Nomia*), famous for the pearly opalescence of their stripes. Sweat bees and alkali bees are important pollinators of fruits and berries as well as seed crops for alfalfa, clover, carrots, marigolds, zinnias, and many more. Most bees in the Halictidae family nest in the ground, though some excavate holes in twigs or rotten wood. The fascinating twig-nester pictured here is a Central American species called *Megalopta genalis*, known for its primitive eusocial behavior and unusual adaptations for flying at night (note the large eyes and ocelli).

ILLUSTRATION © CHRIS SHIELDS.

MELITTIDAE—Oil-Collecting Bees

Members of this small family appear among the oldest known fossil bees, and most taxonomists consider them relics of an ancient lineage. Many of the roughly 200 species are highly specialized, collecting pollen from only one or a very few varieties of flowers. Two genera (*Rediviva* and *Macropis*) share the habit of gathering droplets of oil from the flowers they visit. They use this unusual harvest to line their nest cells, and also as a supplemental food for their larvae. In southern Africa, oil gathering has led one group of related species to evolve front legs that stretch up to twice as long as their bodies, such as in the *Rediviva longimanus* pictured here. These ungainly appendages help the bees probe for oils deep within flowers called twinspurs. The relationship is coevolutionary, with the two spurs of the flowers perfectly sized to receive the legs of the bees. Mellitids are generally solitary and nest in the ground or in rotten wood. ILLUSTRATION © CHRIS SHIELDS.

MEGACHILIDAE—Leafcutter Bees, Mason Bees, and Wool-Carder Bees

Members of this large and widespread family (> 4,000 species) share the charming trait of carrying pollen on what amounts to their bellies. Most groups also incorporate distinctive building materials into their nests. Mason bees (*Osmia*) dab on bits of mud or clay, and wool-carders (*Anthidium*) make a felt of plant hairs. Other groups glue together pebbles or flower petals, while leafcutters (*Megachile*) use their powerful mandibles to snip and assemble strips of vegetation. They are highly efficient pollinators, and several species can be purchased commercially for the pollination of fruit trees, alfalfa, and almonds. This family includes the largest bee in the world, Wallace's giant bee (*Megachile pluto*), whose wingspan exceeds 2.5 inches (63.5 millimeters). Naturalist Alfred Russel Wallace discovered a single specimen in 1859, and few people have seen it since—the bee remains known from only three islands in Indonesia, where it inhabits the nests of a tree-dwelling termite. Members of this family are mostly solitary, though a few (including Wallace's bee) live communally. Wallace's bee (bottom right) and two leafcutters are pictured here. ILLUSTRATIONS © CHRIS SHIELDS.

APIDAE—Bumblebees, Carpenter Bees, Digger Bees, Honeybees, Long-Horned Bees, Orchid Bees, Squash Bees, and Stingless Bees

With over 5,700 described species, Apidae is the largest of all bee families and one that taxonomists have called "enormously diverse" in both the appearance and habits of its members. It contains not only many of our most familiar groups, including bumblebees (*Bombus*) and honeybees (*Apis*), but also scores of lesser-known varieties, from fuzzy blue carpenter bees (*Xylocopa caerulea*) to rainbow-iridescent orchid bees (*Euglossa*) to species with bizarre antennae longer than their bodies (*Eucera*). Members of this family nest everywhere from cliffs to tunnels in the ground to abandoned rodent burrows or hollow trees. Some build their nests from mud (*Euleama*) or plant resins (*Melipona*), while others bore holes in wood (*Xylocopa*) or remove the pith from broken stems and twigs (*Ceratina*). The family contains many solitary species as well as the highly social honeybees and stingless bees (e.g., *Melipona*, *Trigona*), whose complex societies can number in the tens of thousands. Over 30 percent of the species in this family are klepto-parasites, or "cuckoo bees," which do not build nests or gather pollen, but reproduce instead by laying their eggs in the nests of others. (This highly successful lifestyle occurs in most bee families and has evolved independently over twenty times.) The oldest known fossil bee (*Cretrigona*) looks very much like a modern stingless bee (*Trigona*), and experts believe that the Apidae evolved early, flourishing right alongside the flowering plants they rely on. A bumblebee in the genus *Bombus*, a *Melissodes* longhorn bee, and a stingless bee in the genus *Trigona* are pictured here. ILLUSTRATIONS © CHRIS SHIELDS.

Bombus ♀

Melissodes ♂

Trigona ♀

Appendix B
Bee Conservation

A portion of the proceeds from this book will be donated to help preserve and protect wild bees. To contribute directly to such efforts, and to learn more about how you can help the bees in your own backyard, contact any of the organizations listed below.

The Xerces Society
628 NE Broadway, Suite 200
Portland, OR 97232
USA
Phone: 855–232–6639
www.xerces.org

Bumblebee Conservation Trust
Beta Centre
Stirling University Innovation Park
Stirling
FK9 4NF
United Kingdom
Phone: 01786 594130
www.bumblebeeconservation.org

Buglife
Invertebrate Conservation Trust
Bug House
Ham Lane
Orton Waterville
Peterborough
PE2 5UU
United Kingdom
Phone: 01733 201 210
www.buglife.org.uk

Notes

Introduction: The Buzz About Bees

1 **measurable fear reaction:** See Seligman 1971 for an explanation of the theory, Mobbs et al. 2010 for an experimental example, and Lockwood 2013 for an in-depth exploration of this topic.

1 **synapses associated with disgust:** This reaction to insects appears early in life and is considered a "core" disgust. See Chapman and Anderson 2012 for a fascinating summary of disgust research.

2 **without ever having to actually see one:** The Chinese express perhaps the greatest affection for crickets, keeping them as household pets and even staging elaborate chirping competitions. But while they might be transported or displayed temporarily in bamboo cages, most pet crickets live their lives tucked out of sight in gourds or clay pots (which also serve to amplify their song).

4 **dating back more than 8,500 years:** See Roffet-Salque et al. 2015.

5 **long before they tamed horses:** Pinpointing dates of domestication is tricky business, and often the subject of heated debate. The comparisons in this passage rely on a conservative estimate for beekeeping at 6,500 years ago, halfway between the first possible signs noted in Roffet-Salque et al. 2015 and the advanced techniques practiced by ancient Egyptians. Sources for the livestock and crop dates include Driscoll et al. 2009 and Meyer et al. 2012.

5 **"wheat and the fruit of the tamarisk":** Herodotus 1997, 524.

5 **at least 9,000 years:** To date, the oldest physical evidence of mead or mead-like beverages comes from an analysis of residues found in ancient Chinese jars (McGovern et al. 2004). But honey occasionally ferments in the wild, raising the intriguing possibility that our ancestors stumbled across this idea much earlier.

6 **spiked with narcotic roots and bark:** In addition to mead, honey itself can be intoxicating when bees forage on the nectar of particular narcotic plants. Reports of hallucinogenic honey come from the Mayans as well as from the Gurung people of Nepal and the Ishir of Paraguay, who call a particular class of their shamans "eaters of honey" (Escobar 2007, 217).

6 **over 350 of its 1,000 prescriptions:** Relying on the Syriac *Book of Medicines*, a doctor could readily suggest honey for everything from a sore throat to hiccups, nausea, nosebleeds, heart pain, poor eyesight, or a low sperm count. Wax was a cure-all, too, finding its way into treatments for loose teeth and aching testicles as well as the wounds caused by "swords, spears, arrows, &c." (Budge 1913, CVI).

6 **" . . . impossible to over-estimate their value . . . ":** Ransome 2004, 19.

6 **an impressive 200,000 pounds:** This figure comes to us from Livy's report on a skirmish in 173 BCE, where troops of the Roman praetor C. Cicereius killed 7,000 Corsicans in battle and took another 1,700 prisoner. It marked a doubling of the previous wax tribute enacted after an uprising eight years earlier. Livy's *History of Rome* makes no further mention of the Corsicans. Presumably, they were too busy harvesting wax from their hives to get into much trouble (Livy 1938).

7 **small tablets covered with wax:** Etymologists trace the word "stylus" to the Latin root *sti*, meaning "to prick," the same root that forms the basis of "sting." This raises the charming idea that Roman scribes scribbled away on their beeswax tablets with the linguistic equivalent of "stingers."

8 **Bee-priestesses, known as *melissae*:** Melissa remains a popular woman's name, as does the related Melina, Greek for "honey." In Hebrew, the word for bee is *d'vorah*, source of another familiar name, Deborah.

Chapter One: A Vegetarian Wasp

17 **the benefits of a good patch:** Evidence suggests that sand wasps, like many solitary bees, may also benefit from "safety in numbers," lowering the individual risk of predation or parasitism by clustering their nests together.

19 **something entirely different:** In general, adult wasps eat flower nectar or fruit pulp to fuel their own bodies, while seeking out prey or carrion to provision their larvae.

19 **outnumber their sphecid relations:** Data from O'Neill 2001.

21 **the oldest unequivocal bee:** A putative bee with intriguing, wasp-like traits was described from Burmese amber (Poinar and Danforth 2006), but has since been questioned by several knowledgeable experts. Unfortunately, the specimen remains in private hands and is currently unavailable for reexamination. Fossils from Burmese amber hold great promise, however, as they date to the mid-Cretaceous, 100 million years ago, a critical and entirely undocumented period for bee evolution.

23 **regurgitating it in the nest:** Yellow-faced bees in the genus *Hylaeus* were once considered primitive, in part because of their wasp-like appearance and pollen-swallowing habits. More recent studies suggest they evolved later, becoming wasp-like only after adopting the pollen-swallowing habit. The strategy used by early bees remains very much in debate, but preeminent bee scholar Charles Michener (2007) believed that proto-bees carried pollen externally, in whatever hair they had available.

26 **trapped by its original sticky ooze:** There is a terrific irony in finding insects trapped in amber, since it was often insects (particularly beetles) that wounded the tree and caused the pitch to ooze in the first place. Intended as a defensive mechanism, the resin may or may not have warded off the tree's attackers. But in many cases it did succeed in preserving them, as well as innocent passersby, for all time.

26 **ravages of insect-borne disease:** The sand fly's gut contained an insect-borne protozoan related to those that cause sleeping sickness, Chagas disease, and leishmaniasis. See Poinar and Poinar 2008.

28 **resin to use in making their nests:** The production of resin in flowers occurs in several hundred mostly tropical species. While

it's conceivable this habit originated as a defense against seed or petal-eating herbivores, in all known cases the resin now serves as a reward for pollinators (mostly bees). See Armbruster 1984, Crepet and Nixon 1998, and Fenster et al. 2004.

29 **still fragrant after 44 million years:** Later, Noah and I learned that fossilized resin retains another of its ancient properties: flammability. A small chunk, ignited on a brick in the flowerbed beside my office, burned like fury for several minutes, producing a choking cloud of black smoke. Our experiment confirmed that the Germans were right to call amber *bernstein*, "burn stone," a word sometimes adopted as a surname by workers in the amber trade or people from amber-producing regions.

Chapter Two: The Living Vibrato

31 **" . . . knows not the subject":** Linnaeus attributed this quotation to medieval scholar Isidore of Seville (c. 560–636), who included the idea, in slightly different words, in the first volume of his famed *Etymologiae*.

35 **"asking an ornithologist . . . ":** Dr. Laurence Packer, "An Inordinate Fondness for Bees," n.d., archived at www.yorku.ca/bugsrus /PCYU/DrLaurencePacker, accessed September 5, 2016.

36 **Experts don't agree on the function of the hairs:** Some believe that honeybee eye hairs are *mechanoreceptors*—structures that are physically sensitive to changes in wind direction and air speed. One study famously shaved off the eye hairs from captive bees and found their navigation skills subsequently impaired in windy conditions (as cited in Winston 1987). Other studies describe no perceptible nerve cells at the base of the hairs and point out that hairs tend to wear off as the bees age with no apparent ill effects (e.g., Phillips 1905).

36 **almond reek of potassium cyanide:** Returning home from The Bee Course, I experienced a tense moment that must be familiar to entomologists who travel by plane. As I stood in line for airport security, it suddenly occurred to me that I had two kill jars full of potassium cyanide in my carry-on luggage. I felt like a deer in the headlights watching that bag disappear into the X-ray machine . . . but it passed through without a hitch. I was glad to keep the jars—cyanide is hard to come by. But knowing that my bag contained crudely corked vials of a deadly poison did raise the

discomfiting question of what the passengers around me might be carrying!

37 **more than 8,000 specimens in all:** Keynes (2000) combed Darwin's notes from the *Beagle* voyage and assembled a list of zoological specimens, including 1,529 preserved in spirits of wine, 3,344 in other spirits, and 576 not in spirits. Among the many gems was #1,934, collected in the Falkland Islands—"Teeth of rat out of stomach of a Hawk shot in the country." Porter (2010) reviewed Darwin's plant collections and found 2,700 specimens on 1,476 herbarium sheets at Cambridge, where the bulk of his botanical efforts are stored. Note that these totals do not include Darwin's geological or paleontological specimens, which were also extensive.

37 **Alfred Russel Wallace was even more prolific:** Wallace's hefty inventory included mammals, reptiles, birds, shells, and insects, as reported in his wonderful account titled *The Malay Archipelago* (Wallace 1869, xi). Remarkably, beetles accounted for 83,200 of his specimens, over two-thirds of the total.

38 **lattice of translucent chitin:** For a full explanation of this phenomenon, which is also found in some beetles and butterfly scales, see Berthier 2007.

39 **once blinded a shepherd:** See Graves 1960, 66.

39 **three basic parts: head, thorax, and abdomen:** The unusual positioning and development of the waist in bees, wasps, and ants technically put the first segment of the abdomen onto the thorax. Functionally, this distinction is irrelevant, and most authors simply call the back end of a bee the abdomen (or *metasoma*), as in other insects.

39 **"the wing of the bee . . . ":** Aristotle 1883, 64.

41 **"a blizzard of odors":** Schmidt 2016, 12.

41 **the bee will turn to follow:** Researchers demonstrated this ability with an ingeniously wacky experiment involving a simple, Y-shaped maze. Bees released at the base of the Y could easily locate a scent lure placed in one of the branches. With their antennae painstakingly crossed, however, and held in place by a dab of glue, the same bees invariably followed their reversed antennal signals to the empty branch of the Y (as cited in Winston 1987).

42 **over half a mile (one kilometer) away:** Scent plumes are difficult to measure in the wild, where it can be impossible to tease apart the effects of visual and other cues from scent in terms of how

a foraging bee finds a flower. Jim Ackerman cleverly overcame this challenge by attracting male orchid bees to an island isolated in the middle of Lake Gatun, Panama. Since no orchid bees occurred on the island naturally, all the bees visiting his fragrant baits had to have come from the surrounding forest, lured across half a mile of open water by scent alone (David Roubik, pers. comm.).

42 **landing on angled surfaces:** See Evangelista et al. 2010.

42 **". . . fainted from the pain":** Porter 1883, 1239–1240.

42 **light intensity and polarization:** Ocelli occur in all sorts of arthropods, from insects and spiders to horseshoe crabs. Their capabilities vary and remain mysterious in many cases. For bees, there is mounting evidence that they play a role in navigation during low-light conditions. The few species adapted to crepuscular and nocturnal foraging all have developed greatly enlarged ocelli (see Wellington 1974, Somanathan et al. 2009).

43 **speed, distance, and trajectory:** Bees can also use this ability when they are in motion, and it helps them to judge the distance to nearby stationary objects. Combined with their acute and directional sense of smell, this gives them a rich, three-dimensional perception of their surroundings (see Srinivasan 1992).

43 **yellower shades of orange:** With very few exceptions (e.g., the Japanese honeybee *Apis cerana japonica*), bee eyes lack the optic receptors necessary to distinguish the color red. Many can still locate red flowers, however, by sensing the differences in light intensity generated by red against a green backdrop (see Chittka and Wasser 1997).

43 **more than a quarter of all flowering plants:** For a good discussion of bee purple and other ultraviolet flower phenomena, see Kevan et al. 2001.

45 **the depths of the borage flowers:** Many desert flowers hide their nectar in deep pockets to reduce evaporation. The extraordinary mouthparts of this species allow it to feed while perched on top of a deep flower, where it can continue scanning its surroundings for danger while its tongue reaches the nectar inside (Packer 2005).

45 **a German physicist and a Swiss engineer:** Magnan's assertion is the only one to appear in print, but another version of the story, perhaps apocryphal, traces the famed bumblebee calculations to a cocktail party attended by either Ludwig Prandtl, Jacob Ackeret, or their students.

46 **"women who don't know they can fly":** As quoted in Hershorn 1980.

47 **before the next impulse arrives:** See Heinrich 1979.

47 **further increasing their lift:** For a good review of bee aerodynamics, see Altshuler et al. 2005.

47 **peak of Mount Everest:** In a splendid example of inventive fieldwork, Dillon and Dudley (2014) captured local bumblebees (*Bombus impetuosus*) in the mountains of western China and placed them in flight chambers where air pressure could be reduced to simulate increased elevation. The bees maintained flight not by increasing the frequency of their wingbeats, but by increasing their amplitude (i.e., sweeping the wings wider with each stroke).

50 **tiny holes in the cuticle:** Because bee bodies are small, the respiratory and circulatory systems can be simple, with blood flowing freely through much of the body cavity, exchanging nutrients and waste directly with the cells. Air also diffuses broadly, eliminating the need for fussy lungs and the transport of oxygen via hemoglobin.

51 **rarely sting at all:** Bees suffer from a public relations problem. The vast majority of stings blamed on them actually result from encounters with wasps, particularly the social species in the family Vespidae known as yellow jackets, paper wasps, and hornets. Though fascinating in their own right, these creatures often display a regrettable tendency toward testiness and downright aggression. Even entomologists tread lightly around them, and I once heard a social wasp expert begin a public lecture with the admission, "Nobody likes a social wasp."

52 **group defense tactics:** E. O. Wilson and other evolutionary thinkers maintain that the group defense of a nest is an essential condition for developing a fully eusocial lifestyle, so it's not surprising that the fiercest bee stings are found in highly social species like honeybees. What is surprising is that the largest group of eusocial bees have tiny, reduced stingers incapable of inflicting harm. The Meliponine or "stingless" bees include approximately 500 mostly tropical species. Their evolutionary story remains controversial, but it appears they lost their sting after becoming eusocial, and many have since compensated by developing foul odors, mobbing behaviors, and painful bites augmented with caustic, blistering chemicals. Suicidal, kamikaze-style biters have even been proposed as a measure of altruism for stingless species.

But why they didn't simply retain the stinging ability that helped them achieve sociality in the first place remains anyone's guess (see Wille 1983, Cardinal and Packer 2007, and Shackleton et al. 2015).

53 **"live" apart from the bee:** The sting of the honeybee is truly devilish. In addition to pumping venom, the disembodied stinger actively pushes its lancets deeper into the victim, and emits alarm pheromones that call in sister bees to continue the attack.

53 **". . . father's angry rays":** Maeterlinck 1901, 24–25.

CHAPTER THREE: ALONE TOGETHER

55 **". . . solitude is a beautiful thing":** This quotation, in various forms, is often erroneously attributed to the nineteenth-century author and playwright Honoré de Balzac. But that particular Balzac never said or wrote anything of the kind. The quote comes instead from the pen of Jean-Louis Guez de Balzac (no relation), a prolific seventeenth-century essayist, letter-writer, and early member of the French Academy. Balzac 1854, 280; translation confirmed by S. Rouys, pers. comm.

60 **complete metamorphosis:** The evolution of metamorphosis in insects is poorly understood, although fossils with a distinct larval stage date to at least 280 million years ago. It may offer some advantage in reducing competition between offspring and adults, particularly for those species with long-lived larvae. Regardless, it has become an extremely successful life strategy, accounting for over 80 percent of all insects, including bees, wasps, ants, flies, fleas, beetles, moths, and butterflies.

60 **rouse them from their torpor:** In some bees (including several close relatives of orchard masons), a portion of each season's offspring remains in torpor for an extra year. Theoretically, this delay evolved as a hedge against bad weather, sporadic pollen and nectar resources, or catastrophic events that could wipe out an entire cohort of emerging bees. But the strategy is not without risks—the longer an individual remains dormant in the nest, the longer it is exposed to parasites and pathogens. And any two-year bees positioned near the entrance will be destroyed by the one-year bees chewing their way out from behind (see Torchio and Tepedino 1982).

61 **" . . . real story of the Hymenoptera":** Taking this comment an interesting step further, evidence suggests that the common ancestor of all stinging wasps, ants, and bees was a parasitoid. The larvae of these groups postpone defecation until late in their cycles, a habit that parasitoids use to avoid sickening their hosts too soon. It appears to be a shared trait inherited from a common ancestor, suggesting that the parasitoid lifestyle has been lost (e.g., bees, ants) and regained (e.g., some vespoid wasps) several times over the history of this diverse group.

62 **in other birds' nests:** The term "cuckold" also comes to us from this cuckoo bird habit, an analogy people recognized far earlier than they began talking about the behavior of bees. The first use of "cuckold" in English predates the phrase "cuckoo bee" by nearly six centuries.

63 **sickle-shaped mandibles:** The purpose of these deadly mouthparts is indisputable. They appear only in the earliest life stages of cuckoo bee larvae. Once the host larva is safely dispatched, cuckoo grubs lose their weaponry and develop just like any normal baby bee.

64 **stuffed with bee bread:** The size of an adult bee directly reflects the amount of food it received as a larva. In solitary species like orchard masons, this often leads to visibly larger females, but it can also indicate the proficiency of the mother and the environmental conditions during the flight period. A season with poor weather or scarce floral resources results in smaller adults the following year. In social species like honeybees or bumblebees, the larvae chosen to be raised as queens receive extra food as larvae, triggering fertility as well as a large size. (Honeybees go so far as to produce a particularly nutrient-rich substance, "royal jelly," exclusively for use in the diets of potential queens.)

65 **make their homes in dessicated cow pies:** Cane, 2012, 262–264.

67 **confuse predators at close range:** Debate over the purpose of zebra stripes dates back to arguments between Darwin and Wallace. Recent studies suggest that they help keep zebras cool and also repel biting flies, though other evidence still points to the importance of visual effects (see How and Zanker 2014, Larison et al. 2015).

67 **" . . . big primate in Africa":** E. O. Wilson interview, "E. O. Wilson on the 'Knockout Gene' That Allows Mankind to

Dominate Earth," Big Think, n.d., http://bigthink.com/videos
/edward-o-wilson-on-eusociality.

68 " . . . majesty of law": Virgil 2006, 79.

68 " . . . no ready answer": Michener 2007, 15.

70 "morphologically monotonous": Ibid., 354.

71 particularly close relatives: This relatedness is reduced, however,
when females mate with and store sperm from more than one
male, which is certainly the case in some species.

Chapter Four: A Special Relationship

77 "The botanist should make interest . . . ": Thoreau 2009, 169.

80 "pollen wasps": The most important pollen wasp communities exist in southwestern Africa, where Gess and Gess (2010)
reported ground-nesting aggregations in the thousands for some
species and regular flower visitation to asters, bellflowers, and a
number of other families. In terms of pollination, they're generally considered less important than the various bees attracted to
the same blossoms. But for certain flower species at certain times
of the year, wasp visitors greatly outnumber bees and may be their
most effective pollinators.

80 Winston Churchill coined the phrase: The full text and even
a partial video of Churchill's address can be viewed at the International Churchill Society's website at www.winston
churchill.org/resources/speeches/1946-1963-elder-statesman
/120-the-sinews-of-peace.

80 "coevolutionary vortices": Thompson and other authorities now
use this phrase regularly, but they trace it to the 1984 book *Insects
on Plants* by Donald Strong, John Lawton, and Sir Richard Southwood (Strong et al. 1984).

81 branched and downy as feathers: The connection between bees
and fuzz and pollen is a tight one—alter any aspect of the dance
and things tend to change quickly. Cuckoo bees, for example,
don't collect pollen, and so have little reason to be hairy. Many
of them have therefore dispensed with their fuzz and look quite
smooth and wasp-like, though a careful microscopic inspection
will always reveal a few residual branched hairs still clinging to
the legs, face, or body.

84 intimate players in the pollination process: As regular nectar
feeders, wasps are erratic pollinators of many species, but rarely

serve as a plant's dedicated partner. Notable exceptions include fig wasps, certain pollen wasps with hooked facial hairs, and the males of various species duped into pseudocopulation with orchids.

84 " . . . favoured intercrossing": Darwin 1879, from a facsimile in Friedman 2009.

84 evolve before the Cretaceous: The precise age of angiosperms remains hotly debated, but a combination of fossil and genetic data suggest they may have arisen during the Jurassic period, perhaps eking out a living as tropical forest shrubs before diversifying during the Cretaceous (reviewed in Doyle 2012).

85 "so blue and golden": from "Flowers," Longfellow 1893, 5.

86 perky splashes of red: By one way of thinking, even red would be limited. Some experts argue that birds favor red blossoms less out of preference than opportunity. They visit a range of colors, but since red flowers are invisible to most bees (or at least more difficult to find), they offer birds a nectar source with far fewer competitors, encouraging bird/plant specialization. Without bee competition driving that system, birds might be just as likely to visit other colors, such as in the bee-free flora of the Juan Fernández Islands, where only three of fourteen hummingbird-pollinated species are red.

86 "odorous at sunrise": From "Give Me the Splendid Silent Sun," Whitman (1855) 1976, 250.

87 " . . . dorsal surface of the humble-bee": Sutherland 1990, 843.

88 "What a proboscis . . . ": From a letter to Hooker dated January 30, 1862, archived at the Darwin Correspondence Project, University of Cambridge, www.darwinproject.ac.uk. See also Kritsky 1991.

89 leaky, worm-eaten ship: Selkirk's instincts about the vessel proved correct. The Cinque Ports foundered three months later off the coast of Colombia. Its captain and surviving crewmembers were captured and imprisoned by Spanish colonial authorities.

89 small, round, and greenish white: In a thorough and fascinating review of the Juan Fernández flora, Bernardello et al. 2001 found that 73 percent of native flowers are white, green, or brown. Only 12 percent of the blooms are yellow, and blue, the most bee-specific shade, accounts for just 5 percent of flowers. Similarly, over 75 percent of the flowers are round or inconspicuous, with just 2 percent showing the bilateral flags or other symmetry common in bee-specific blooms.

89 **better accommodate hummingbird bills:** In at least one species, the color also shifted from pure blue (a bee color) to a purple that may be more attractive to the birds (see Sun et al. 1996).

91 **"All bees have them . . . ":** Brady's statement relies on the logic of shared inheritance, a key principle in the study of evolution. When a group of related organisms share a trait, like branched hairs, the simplest explanation is that the trait was inherited from a common ancestor rather than reinvented over and over again individually.

91 **shifted from bumblebees to hummingbirds:** For more on these fascinating studies, see Schemske and Bradshaw 1999 and Bradshaw and Schemske 2003.

91 **trade in bees for hawk moths:** See Hoballah et al. 2007.

92 **not nearly as concentrated:** In experimental settings, bees consistently choose nectar with the highest sugar concentration available (e.g., Cnaani et al. 2006). This behavior is easily observed in the wild, where bees quickly learn to visit hummingbird feeders and other reliable sources of sweetness. At the La Selva Biological Station in Costa Rica, for example, the best place to watch stingless bees in the genus *Trigona* is on the porch of the cafeteria, where they line up to feed from the rims of bottles of the ubiquitous local condiment, Lizano Sauce.

92 **bees make honey:** At more than 80 percent sugar content, honey is roughly twice as sweet as the nectar of the average bee-pollinated flower.

93 **play a role in their mating rituals:** Exactly how male Euglossines use these perfumes remains unclear, but it's thought they may assist in establishing sites called *leks*, where multiple males gather to vie for females. One thing is certain—the scents alone don't attract females, since females never visit the flowers.

95 **" . . . I cannot conjecture":** Darwin 1877, 56.

99 **begets new species:** For one of many fascinating studies of this process in *Ophrys*, see Breitkopf et al. 2015.

99 **tongue length and the depth of flower spurs:** The trend toward longer tongue length ranks among the strongest evolutionary signals in bees and presumably resulted from a continuing need to reach inside deepening flower spurs. For plants, developing spurs attracts a narrower, more dedicated group of pollinators and apparently leads

to diversification. Spurred lineages like monkshood (*Aconitum* spp.) and columbine (*Alquilegia* spp.), for example, both contain dozens of species, while a close relation, love-in-the-mist (*Nigella* spp.), lacks spurs and includes only a handful of species.

99   **corresponding risks of dependence:** Many plants in specialized pollinator relationships hedge against the downside by retaining some level of self-fertility, the ability to produce seed from their own pollen. In fact, this failsafe may contribute to the relative evolutionary ease with which many flowers appear to be able to experiment with new scents, colors, and other pollination traits.

Chapter Five: Where Flowers Bloom

103   **"Supply creates its own demand":** This phrase is the common summary of a principle put forth by French economist Jean-Baptiste Say in his book A *Treatise on Political Economy*, first published in 1803.

104   **US Department of Agriculture Bee Lab:** Although it is known colloquially as the Bee Lab, this excellent facility officially goes by a much longer title: the United States Department of Agriculture Pollinating Insect–Biology, Management, Systematics Research unit.

104   **bees in the genus *Anthophora*:** The genus name for digger bees comes from Greek words meaning "flower-bearers." But for this species the real story lies in their specific epithet, *bomboides*, a reference to the bumblebee genus, *Bombus*. The result is an uncommonly clear scientific title: *Anthophora bomboides*, the digger bee that looks like a bumblebee.

105   **" . . . precipitous earthy banks":** Fabre 1915, 228.

106   **" . . . pursued most industriously":** Nininger 1920, 135.

108   **classic evolutionary bluff:** Known as Batesian mimicry, this strategy involves a harmless species adopting the warning coloration of a toxic, stinging, or otherwise dangerous one. The former therefore benefits by exploiting the honest signal of the latter, warding off potential enemies or predators. This form of mimicry is named for nineteenth-century English explorer and naturalist Henry Walter Bates, who first described it in various Amazonian butterflies.

108 **primary means of defense:** While *Anthophora bomboides* no longer sting, they do show some evidence of developing other defensive behaviors. During one of my visits to the cliff while working on this chapter, a female got tangled in my net, and it took me some time to remove her. Soon after, I noticed a number of other individuals hovering around me, darting in close before backing off again. In all previous visits the bees had ignored my presence completely, but now a dozen or more were badgering me—they even followed me out onto the sand of the beach. Had the trapped bee released an alarm pheromone in the net? To test this idea I walked far down along the cliff and extended the net into a different part of the colony. It was immediately surrounded by hovering bees. As a solitary species, diggers have no history of coordinated defense, but they are gregarious and do occasionally share nest tunnels. Defense lies at the heart of definitions for social evolution—Could this nascent aggression be a start down that path? No expert I talked to had any explanation, but Brooks (1983) noticed the same behavior in *A. bomboides*, and Thorp (1969) saw something similar in another *Anthophora* species. It's a wonderful thesis topic just waiting for the right graduate student!

108 **" . . . cannot be induced to sting":** Brooks 1983, 1.

108 **" . . . chimney of clay":** Nininger 1920, 135.

108 **brimming with nectar:** Digger bees also transport water in their honey crops, using it to moisten the soil as they sculpt their tunnels, chambers, and turrets. During the peak of nest construction, females make as many as eighty round-trip flights to freshwater sources every day (Brooks 1983).

111 **pretty much anywhere:** Collapsible insect nets can also be tucked out of sight quickly, a handy trait in places where collecting might not be welcomed. Entomologists have been known to call them "National Park Specials."

116 **"If you build it, they will come":** This common expression is a slight misquote from the 1989 movie *Field of Dreams*, which featured an Iowa farmer constructing a baseball diamond in his cornfield after hearing a voice that whispered, "If you build it, *he* will come."

118 **untripped and unfertilized:** The honeybee habit of robbing nectar from alfalfa flowers sets the stage for even higher levels of thievery. During our tour through the valley, Mark showed us where a commercial beekeeper had set up shop on a rented

lot surrounded by alfalfa fields. Scores of busy hives crowded the small space, undoubtedly rich with honey. But since those bees failed to pollinate most of the flowers they visited, the practice amounted to piracy—draining flowers of their nectar while providing nothing in return, thereby reducing the farmers' seed set, yields, and profits. "It's not that I don't like honeybees," Mark explained, a little gruffly. "I just don't like beekeepers."

Chapter Six: Of Honeyguides and Hominins

123 **Honeyguides and Hominins:** In modern usage, the word "hominin" refers to a specific primate subgroup that includes our genus, *Homo*, and its extinct relations, including *Australopithecus* and *Ardipithecus*. It is often confused with "hominid," a family-level category that includes all of the great apes—the hominins plus chimpanzees, gorillas, and orangutans (family Hominidae). (The words were interchangeable in older versions of primate taxonomy, which put the other great apes in a different family.) Like pretty much everything else in human paleontology, however, these definitions are still debated. Some experts now prefer to lump chimpanzees, our closest living relations, in with the hominins.

123 **"No bees, no honey":** This is the common translation of the proverb recorded by Erasmus in Latin as "Neque Mel, neque Apes." Bland 1814, 137.

124 **impacts on local species:** Cane and Tepedino (2016) argued that the most significant impacts of honeybees on native North American species come not in agricultural or developed areas, but in wild habitats, particularly in the American West, where commercial honeybee hives are often "pastured" for several months after the bees have pollinated various crops.

126 **". . . in hopes of a better salary":** Sparrman 1777, 44.

133 **exclusively on people:** More evidence for the bird's strong connection to people can be found in its absence. Near cities, towns, and farming settlements, where hardly anyone still hunts for honey, the birds have begun to lose their guiding habit. Some conservationists now call for the reintroduction of traditional honey hunting to Africa's national parks, in an attempt to preserve not just the honeyguide, but the behavior that defines it.

133 **power in the form of glucose:** Under starvation conditions, when glucose becomes limited or unavailable, the brain can run

for short periods of time on ketones derived from the breakdown of fatty acids.

134 **specimen from the genus *Australopithecus*:** Some authorities now place Nutcracker Man in a separate genus, *Paranthropus*, the "robust australopithicenes." It is also occasionally referred to as *Zinjanthropus*, the title originally proposed by the Leakey family. Naming disputes aside, experts generally agree that it wasn't a direct human ancestor, but rather, one of several closely related hominins that inhabited East Africa around the time that the genus *Homo* evolved.

135 **dawn of the Neolithic:** Bernardini et al. (2012) and Roffet-Salque et al. (2015) provide good evidence for honey use in the Neolithic.

135 **juicy prehistoric smooch:** Though the kissing idea grabbed all the headlines, this study revealed great insights into the Neanderthal diet—in particular, how it differed from place to place, taking advantage of local offerings ranging from woolly rhinoceroses to wild sheep, mushrooms, pine nuts, and moss. Because the authors analyzed traces of DNA rather than chemical signatures, however, they were unable to look for evidence of honey (Weyrich et al. 2017).

135 **cooperation and sharing:** As with meat, fruit, tubers, and other foraged foodstuffs, honey is shared widely among the Hadza. But because it is a particularly favored item, it can also be the object of deceit. When there wasn't enough to go around, Alyssa often observed hunters hiding pieces of comb under their shirts to give to their wives and children.

135 **as regularly as the Hadza:** The Hadza tribe's honey habit is not an isolated example. Honey crops up as an important food source for hunter-gatherers in virtually every landscape that contains honey-producing bees. The Mbuti people of the Ituri rainforest in Congo, for example, also rank the products of the hive as their favorite foods. They raid the nests of at least ten different bee species and rely on honey, pollen, and larvae for 80 percent of their calories during their annual "honey season," a period of mass flowering and high bee abundance that lasts for up to two months (see Ichikawa 1981).

136 **". . . outcompete other species":** Crittenden 2011, 266.

139 **". . . honey bright as gold":** Brine 1883, 145.

139 **". . . bloom, it is the bumblebee":** Stableton 1908, 22.

Chapter Seven: Keeping Dumbledores

141 " . . . like directing the sunbeams": Thoreau 1843, 452.

143 "the rounded end of a wooden stick . . . ": Sladen 1912, 125. I've tried this, and the end of a wooden spoon works quite well. An added bonus: your kitchen will be filled with the rich smell of molten beeswax while you work.

146 "dying, queenless hive": Tolstoy (1867) 1994, 998.

146 " . . . criminal world of London": Doyle 1917, 302.

146 beekeeping canon: Among the many scores of beekeeping books, standouts include Sue Hubbell's memoir *A Book of Bees* (1988); *The Queen Must Die* (1985), by William Longgood; and *The Bee-keepers Bible* (2010), a how-to volume by Richard Jones and Sharon Sweeney-Lynch.

146 references of a different variety: Plath clearly knew about other kinds of bees too. In one poem she describes peering into the pencil-thin holes of a ground-nesting bee in a way that could only come from personal experience. Someday, an English major with an entomological bent will write a great thesis correcting all the misconstrued literary interpretations of Plath's bee metaphors. When she referred to a solitary bee, for example, she clearly wasn't talking about a honeybee that happened to be alone!

147 harmlessly in his fist: Plath 1979, 311.

147 boot stuffed full of twigs: Although this wren got the best of our bee, the shoe is sometimes on the other foot. Several studies have noted bumblebee queens evicting birds from nest boxes, in some cases even after the birds had begun laying eggs. Playback experiments on two species of tits in Korea showed that the sound of buzzing was enough to make many incubating females flee the nest (Jablonski et al. 2013).

152 " . . . Bumble Bee's heart": Coleridge 1853, 53.

153 species known as the fuzzy-horned bumblebee: The scientific names of the bees in question are *Bombus sitkensis* and *Bombus mixtus*. Because few casual observers know the differences among the world's 250 bumblebee species, many of them lacked common names until very recently. Jonathan Koch, lead author of a field guide to the bumblebees of Western North America, found himself inventing names right up until the book went to press. He told me in an email that *B. mixtus* earned the title "fuzzy-horned bumblebee" because the males have telltale tufts of orange fuzz on

the inner faces of their antennae. And also because it "just sounds cute."

154 **pollinators for things like red clover:** In *On the Origin of Species*, Darwin proclaimed bumblebees to be the only pollinators of red clover, but he later learned that honeybees also visit the flowers (as do a variety of solitary bees). He was mortified by his error and wrote to a friend, "I hate myself, I hate clover, and I hate bees" (from a letter to John Lubbock, September 3, 1862).

154 **" . . . flowers in that district!":** Darwin 1859, 77.

Chapter Eight: Every Third Bite

158 **"Two all-beef patties, special sauce . . . ":** The components of a Big Mac do vary slightly in different corners of the globe. South Africans add a slice of tomato, for example, and in India, where cows are revered, the beef is replaced with chicken or lamb.

159 **feedlots have been known to fatten up their charges:** Feeding candy and other odd things to cows is a common practice, particularly when grain prices are high (see Smith 2012).

160 **Canola—a trade name for field mustard:** The common name for this oil-producing field mustard is "rape." To overcome the obvious marketing limitations of that title, crop researchers at the University of Manitoba named their variety Canola, from **Cana**dian **o**il, **l**ow **a**cid.

160 **do visit lettuce flowers:** Among surprisingly few studies of lettuce pollination, Jones (1927) found that bees helped move pollen within and among flowers of the same plants, improving both the rate of fertilization and the number of pollen grains delivered per flower. D'Andrea et al. (2008) used genetic tools to confirm occasional cross pollination, presumably by bees, at distances of up to 130 feet (40 meters), the farthest interval examined.

166 **trees rely on wind for pollination:** The gross inefficiency of wind pollination in date palms has led some experts to suggest that they may have once relied, in part at least, on insects. Which wild palm they descend from remains unknown, but pollination by bees, beetles, or flies is far more common in the family than wind pollination. Also, tissues in the female flowers still appear to be capable of nectar production, and some male varieties produce fragrant blossoms. Brian Brown told me he sometimes sees bees

on the male flowers so covered by the copious pollen that they look "punch drunk." See Henderson 1986, Barfod et al. 2011.

166 **hand-pollinating dates:** Oddly, this intimate practical knowledge did not translate into an inkling of scientific understanding about pollination until well into the eighteenth century. The specifics of pollination, and particularly the role of insects, remained unsettled until Charles Darwin and his contemporaries examined the question in earnest in the 1860s.

167 **staple fruit of the ancient world:** As if bent on putting buzzing insects out of a job, date purveyors of the ancient world used their human-pollinated fruits to usurp another role typically reserved for bees—the production of honey. In the ancient world, "date-honey" was often pawned off as the real thing, or sold as a cheap substitute where bees were scarce. Now referred to as *rub* in Arabic or *silan* in Hebrew, it remains a common sweetener in cuisines throughout the Middle East and North Africa.

167 **"When the male palm is in flower . . . ":** Theophrastus 1916, 155.

169 **relies primarily on hand-pollination:** The orchids that produce vanilla beans normally rely on particular tropical bees but can be easily hand-pollinated with a toothpick. When people figured that trick out in the early nineteenth century, production shifted from Mexico (where the orchids and their associated bees are native) to sites throughout the tropics, depriving Mexican growers of what had been a lucrative vanilla monopoly.

Chapter Nine: Empty Nests

175 **" . . . not stop questioning":** Miller 1955, 64.

176 **identified the species as *Bombus californicus*:** Genetic evidence suggests that *B. californicus* may simply be a local color morph of the more widespread yellow bumblebee *B. fervidus*.

178 **CNN segment called "The Old Man and the Bee":** This is a terrific short feature on Robbin Thorp's search for the Franklin's bumblebee, archived online at www.cnn.com/videos/world/2016/12/08/vanishing-sixth-mass-extinction-domesticated-bees-sutter-mg-orig.cnn/video/playlists/vanishing-mass-extinction-playlist.

180 **emperor's favorite melon:** Though it's often reported that Tiberius had a penchant for cucumbers (*Cucumis sativus*), there is no evidence that cucumbers existed in Europe until medieval times. The related fruit available to him, and one more likely to be favored daily, was the species *Cucumis melo*, ancestor to a range of sweet muskmelons including cantaloupe, honeydew, and casaba (Paris and Janick 2008).

180 **" . . . not supplied with it":** As quoted in Paris and Janick 2008, from the translation by H. Rakham.

183 **white-rumped bumblebee:** The white-rumped bumblebee (*B. moderatus*) lives in Alaska and northern Canada, where populations of the western bumblebee also appear to be stable in spite of the presence of the *Nosema* pathogen. Jamie Strange and other researchers are eager to know whether it's a different strain of *Nosema*, or if climatic or other environmental conditions may be altering its effects.

187 **50, 100, or even 200 square miles:** The best answer to the question of how far honeybees fly is, "As far as they need to." Foraging ranges vary widely, directly reflecting the availability of flowers. A typical distance might be 2 miles (3.2 kilometers), but in landscapes (or at certain times of the year) with sparse blooms, workers regularly travel much farther to find nectar and pollen. An ingenious 1933 study documented bees flying 8.5 miles (13.6 kilometers) from sweet clover fields to hives isolated in an adjacent scrubland (Eckert 1933). Colonies didn't prosper under those conditions, but the experiment showed how far workers are capable of roaming when the need arises. A modern investigation confirmed this finding by decoding the waggle-dances of bees foraging in the moors of Yorkshire, where individuals traveled as far as 9 miles (14.4 kilometers) to reach patches of blooming heather (Beekman and Ratnieks 2000).

187 **sparked heated debate:** What started as scientific uncertainty and disagreement about CCD soon spilled over into the public sphere, with particular acrimony over the possible roles of pesticides and genetically modified crops. Varying results and interpretations have allowed all sides to find at least some evidence to back up their positions, heightening and prolonging the sense of contention. In fact, the CCD controversy itself has now become a subject of academic interest. Social scientists regard its

combination of competing interests, high emotion, and important policy implications as a case study in the public perception of science (see, e.g., Watson and Stallins 2016).

191 **any defensive chemicals at all:** The presence of alkaloids and other defensive toxins in nectar is unusual but widespread, occurring in at least a dozen plant families. The phenomenon remains poorly studied, but it may help forge specialized pollinator relationships. The death camas bee (*Andrena astragali*), for instance, appears to be able to detoxify a potent alkaloid in the nectar and pollen of the plant for which it is named. No other known pollinator can do so. I once found a cuckoo bee that had sipped from a death camas flower; it was so drowsy that I carried it around on my finger for half an hour before finally depositing it, still stunned, on a different (nontoxic) plant. For more information on toxic pollen, see Baker and Baker 1975 and Adler 2000.

194 **cocktails of chemicals:** There is some indication that bees may recognize at least some of the most harmful chemical combinations. Researchers have recently noticed an increase in "entombed pollen," cells filled with pollen that have been abandoned and capped with propolis in the way bees will isolate foreign objects in a hive. Entombed pollen is often oddly colored and tests high in particular fungicides and other pesticides. See vanEngelsdorp et al. 2009.

Chapter Ten: A Day in the Sun

201 **"In this flowery wilderness . . . ":** Muir 1882b, 390.

202 **" . . . vacuum up the nuts":** I later spoke with a technician at a company that makes almond harvesters, who clarified that most models combined vacuuming with a scooping motion to pick up the nuts. Either way, he confirmed that keeping the orchard floor clean and bare was essential for efficient operation.

203 **extinct California butterfly:** The Xerces blue (*Glaucopsyche xerces*) lived exclusively in coastal sand dunes near San Francisco, feeding on native lupines and lotus. It disappeared in the 1940s as a result of habitat loss and is considered the first North American butterfly driven to extinction by human activity.

207 **flowers benefit honeybees, too:** Later that day we passed a huge sunflower field in full bloom, with honeybee hives spaced at

regular intervals around its perimeter. When we slowed down for a closer look, we noticed large cans of sugar syrup perched on top of every hive. It seemed incredible—here, at the height of summer in the midst of rich farmland, the bees needed supplemental feeding to keep them alive. Recalling the prolific, honey-filled hives of his youth in North Dakota, Eric was shocked and a little offended. "It's like looking at a starving, three-legged cow," he said, and we drove on.

208 " . . . a hundred flowers at every step": Muir 1882a, 222.
209 "season of rest and sleep": Ibid., 224.

Conclusion: The Bee-Loud Glade

213 "When summer gluts the golden bees . . . ": Yeats 1997, 15.
215 " . . . And I shall have some peace there . . . ": Ibid., 35.

Appendix A: Bee Families of the World

218 while *still mounted* by the males: See Houston 1984.
220 waiting for up to three years: See Danforth 1999.
224 "enormously diverse": See Danforth et al. 2013.

Glossary

abdomen General term for the rear body segment of a bee or other insect.

antennae Long sensory structures on the head of a bee, fine-tuned to detect everything from smell and taste to airflow, temperature, and humidity.

arthropods A large taxonomic group that includes insects, crustaceans, spiders, and other invertebrates with bodies surrounded by an exoskeleton.

bilateral symmetry Literally, "two-sided" symmetry, a design where something divided along a specific axis forms two halves that mirror one another; common in the shapes of many bee-pollinated flowers.

bumblebee Any of the approximately 250 species of the genus *Bombus*, social bees familiar for their large, fuzzy bodies often striped with bright orange or yellow. Also known colloquially as "humblebees" or "dumbledores."

buzz pollination See *sonication*.

callow An adult bumblebee newly emerged from its cocoon.

cell A single chamber within the nest of a solitary bee that is provisioned with enough pollen and nectar to meet the needs of one larva. This term can also refer to a single compartment within the comb of a highly social species such as a honeybees or a stingless bee.

chitin A naturally tough, fibrous substance made from long chains of sugar molecules that constitutes the main component of arthropod exoskeletons.

251

cuckoo bees   Any of the large number of parasitic bees that lay their eggs in other bees' nests, or, in the case of certain highly social species, kill the queen and take over the workers as their own.

cuticle   The hard, protective outermost layer of a bee's exoskeleton.

digger bees   Bees in the genus *Anthophora*, a large group of robust, hairy bees that sometimes nest in dense aggregations in the ground or in exposed banks and cliffs.

dumbledore   An archaic term for the bumblebee.

endemic   Native to, and exclusively known from, a discrete geographic location.

fructose   A form of sugar found in fruit and honey.

genus   A taxonomic grouping of closely related species all descended from a single common ancestor.

glucose   A form of sugar used to fuel cell activity; it's particularly important for brain function, and particularly abundant in honey.

hibernaculum   An overwintering site for a queen bumblebee, typically a shallow burrow dug into level or sloping ground, often under moss or leaf litter.

hominin   A specific primate subgroup that includes our genus, *Homo*, and its extinct relations, including *Australopithecus* and *Ardipithecus*.

invertebrate   Literally, "without a backbone," a general term applied to arthropods as well as other multicellular creatures that lack spines, from worms and snails to jellyfish.

kleptoparasitism   A form of parasitism involving the appropriation of food or other resources. It is the most common form of parasitism in bees.

leafcutter bees   Members of the genus *Megachile*, who craft their nest cells from strips and disks of greenery snipped from leaves.

lipids   A group of related molecules that include various fats and waxes as well as fat-soluble vitamins.

mandibles   Paired mouthparts of bees used for gripping, biting, crushing, and cutting.

mason bees   Members of several genera in the bee family Megachilidae, who build their nest cells from mud, sometimes mixed with pebbles, sand, or plant materials.

metamorphosis   The complete body transformation from a distinct larval stage to the adult form, a common but not universal developmental transition in insects.

metasoma   A technical term for what is generally referred to as the abdomen of a bee.

microsporidian   A microscopic fungus, or fungal-like organism, that reproduces via spores. In bees, the common pathogen *Nosema* is a microsporidian.

mining bees   Ground-nesting solitary bees in the genus *Andrena*, or, more generally, any bee that excavates a nest tunnel in the ground.

mutation   A random change in the genetic code of an organism; one of the major sources of variability in nature.

mutualism   A relationship between two species that is beneficial to both participants.

Neolithic   The "new stone age," a period of human history generally defined by the advent of farming, animal husbandry, and polished stone tools.

neonicotinoid ("neonics")   A controversial class of pesticides chemically related to nicotine that attack the nervous systems of insects. For bees, they cause death at high doses and a range of sub-lethal effects in smaller amounts.

ocelli   Light-sensitive organs that appear as three translucent domes on the head of a bee that aid in orientation and navigation.

orchid bee   A group of solitary or primitively social tropical species related to bumblebees and honeybees and known for their often elaborate ways of pollinating various orchids.

ovipositor   The egg-laying device at the back end of various insects. In bees and related wasps, this structure has been modified into a stinger.

parasitoid   A parasite whose larvae develop inside the body of another organism, consuming and eventually killing the host. Most bees suffer the attacks of various parasitoid wasp species.

pheromone   A chemical compound used to convey information through the sense of smell; essential for communication among bees.

pollination syndrome   A set of traits in flowers that generally attract a particular suite of pollinators.

poricidal anther   A type of anther where the pollen is trapped inside a central chamber accessible only by a small hole (pore) at one end. Such anthers are common in the nightshade and heath families.

propolis   A resinous substance collected from plant buds by honeybees and some stingless bees for use in hive construction. Also called "bee glue."

protozoan Any of a large group of single-celled organisms that have traditionally included amoebas, flagellates, and cilliates.

radiation In evolution, the sudden and rapid diversification of forms from a single common ancestor.

reproductive isolation A condition where breeding between related populations is prevented by physical or environmental barriers, often considered an important step toward the evolution of distinct varieties and species.

scopa Concentrations of plumose hairs on the legs or abdomens of bees used for the efficient transportation of pollen.

sonication The process of vibrating wings to produce a high-frequency tone that helps shake pollen loose from flowers, particularly those with poricidal anthers. Also called "buzz pollination." See also *poricidal anther*.

speciation The formation of new species.

stamen A "male" structure within a flower that is topped by pollen-filled chambers called "anthers."

sucrose A form of sugar that is a combination of fructose and glucose. Often refined from sugarcane or beets to produce common table sugar.

surfactant A compound that lowers the surface tension of a liquid. Surfactants are commonly used in detergents and are also added to liquid pesticides to increase their efficiency.

taxonomy A branch of science devoted to understanding evolutionary relationships and characterized by the classification and naming of species.

terpene Any of a large group of volatile chemical compounds produced by plants and often used as a defense against herbivory.

thorax The middle body section of a bee or other insect characterized by large muscles to operate the legs and wings.

zygomorphic A botanical term applied to bilaterally symmetrical flowers, such as snapdragons and orchids. Zygomorphic blooms are often bee-pollinated.

Bibliography

Adler, L. S. 2000. The ecological significance of toxic nectar. *Oikos* 91: 409–420.

Alford, D. V. 1969. A study of the hibernation of bumblebees (Hymenoptera: Bombidae) in Southern England. *Journal of Animal Ecology* 38: 149–170.

Allen, T., S. Cameron, R. McGinley, and B. Heinrich. 1978. The role of workers and new queens in the ergonomics of a bumblebee colony (Hymenoptera: Apoidea). *Journal of the Kansas Entomological Society* 51: 329–342.

Altshuler, D. L., W. B. Dickson, J. T. Vance, S. R. Roberts, et al. 2005. Short-amplitude high-frequency wing strokes determine the aerodynamics of honeybee flight. *Proceedings of the National Academy of Sciences* 102: 18213–18218.

Ames, O. 1937. Pollination of orchids through pseudocopulation. *Botanical Museum Leaflets* 5: 1–29.

Aristotle. 1883. *History of Animals*. Translated by R. Cresswell. London: George Bell and Sons.

Armbruster, W. S. 1984. The role of resin in angiosperm pollination: Ecological and chemical considerations. *American Journal of Botany* 71: 1149–1160.

Baker, H. G., and I. Baker. 1975. Studies of nectar-constitution and pollinator-plant coevolution. Pp. 100–140 in L. E. Gilbert and P. H. Raven, eds., *Coevolution of Animals and Plants*. Austin: University of Texas Press.

Balzac, J.-L. G. de. 1854. *Oeuvres*, vol. 2. Paris: Jacques Lecoffre.

Barfod, A., M. Hagen, and F. Borchsenius. 2011. Twenty-five years of progress in understanding pollination mechanisms in palms (Arecaceae). *Annals of Botany* 108: 1503–1516.

Beekman, M., and F. L. W. Ratnieks. 2000. Long-range foraging by the honey-bee, *Apis mellifera* L. *Functional Ecology* 14: 490–496.

Bernardello, G., G. J. Anderson, T. F. Stuessy, and D. J. Crawford. 2001. A survey of floral traits, breeding systems, floral visitors, and pollination systems of the angiosperms of the Juan Fernández Islands (Chile). *Botanical Review* 67: 255–308.

Bernardini F., C. Tuniz, A. Coppa, L. Mancini, et al. 2012. Beeswax as dental filling on a Neolithic human tooth. *PLoS ONE* 7: e44904. https://doi.org/10.1371/journal.pone.0044904.

Bernhardt, P., R. Edens-Meier, D. Jocsun, J. Zweck, et al. 2016. Comparative floral ecology of bicolor and concolor morphs of Viola pedata (Violaceae) following controlled burns. *Journal of Pollination Ecology* 19: 57–70.

Berthier, S. 2007. *Iridescences: The Physical Color of Insects*. New York: Springer.

Bland, R. 1814. *Proverbs, Chiefly Taken from the Adagia by Erasmus*. London: T. Egerton, Military Library, Whitehall.

Boyden, T. 1982. The pollination biology of *Calypso bulbosa* var. *americana* (Orchidaceae): Initial deception of bumblebee visitors. *Oecologica* 55: 178–184.

Bradshaw, H. D., Jr., and D. W. Schemske. 2003. Allele substitution at a flower colour locus produces a pollinator shift in monkeyflowers. *Nature* 426: 176–178.

Brady, S. G., S. Sipes, A. Pearson, and B. N. Danforth. 2006. Recent and simultaneous origins of eusociality in halictid bees. *Proceedings of the Royal Society B* 273: 1643–1649.

Breitkopf, H., R. E. Onstein, D. Cafasso, P. M. Schülter, et al. 2015. Multiple shifts to different pollinators fuelled rapid diversification in sexually deceptive *Ophrys* orchids. *New Phytologist* 207: 377–389.

Brine, M. D. 1883. *Jingles and Joys for Wee Girls and Boys*. New York: Cassel and Company.

Brooks, R. W. 1983. *Systematics and Bionomics of Anthophora—The Bomboides Group and Species Groups of the New World (Hymenoptera—Apoidea, Anthophoridae)*. University of California Publications in Entomology, vol. 98, 86 pp.

Buchmann, S. L., and G. P. Nabhan. 1997. *The Forgotten Pollinators.* Washington, DC: Island Press.

Budge, E. A. W., trans. 1913. *Syrian Anatomy, Pathology, and Therapeutics; or, "The Book of Medicines,"* vol. 1. London: Oxford University Press.

Burkle, L. A., J. C. Marlin, and T. M. Knight. 2013. Plant-pollinator interactions over 120 years: Loss of species, co-occurrence, and function. *Science* 339: 1611–1615.

Cameron, S. A. 1989. Temporal patterns of division of labor among workers in the primitively eusocial bumble bee *Bombus griseocollis* (Hymenoptera: Apidae). *Ethology* 80: 137–151.

Cameron, S. A., H. C. Lim, J. D. Lozier, M. A. Duennes, et al. 2016. Test of the invasive pathogen hypothesis of bumble bee decline in North America. *Proceedings of the National Academy of Sciences* 113: 4386–4391.

Cameron, S. A., J. D. Lozier, J. P. Strange, J. B. Koch, et al. 2011. Patterns of widespread decline in North American bumble bees. *Proceedings of the National Academy of Sciences* 108: 662–667.

Cane, J. H. 2008. A native ground-nesting bee (*Nomia melanderi*) sustainably managed to pollinate alfalfa across an intensively agricultural landscape. *Apidologie* 39: 315–323.

———. 2012. Dung pat nesting by the solitary bee, *Osmia (Acanthosmioides) integra* (Megachilidae: Apiformes). *Journal of the Kansas Entomological Society* 85: 262–264.

Cane, J. H., and V. J. Tepedino. 2016. Gauging the effect of honey bee pollen collection on native bee communities. *Conservation Letters* 10. https://doi.org/10.1111/conl.12263.

Cappellari, S. C., H. Schaefer, and C. C. Davis. 2013. Evolution: Pollen or pollinators—Which came first? *Current Biology* 23: R316–R318.

Cardinal, S., and B. N. Danforth. 2011. The antiquity and evolutionary history of social behavior in bees. *PLoS ONE* 6: e21086. https://doi.org/10.1371/journal.pone.0021086.

———. 2013. Bees diversified in the age of eudicots. *Proceedings of the Royal Society B* 280: 1–9.

Cardinal, S., and L. Packer. 2007. Phylogenetic analysis of the corbiculate Apinae based on morphology of the sting apparatus (Hymenoptera: Apidae). *Cladistics* 23: 99–118.

Carreck, N., T. Beasley, and R. Keynes. 2009. Charles Darwin, cats, mice, bumble bees, and clover. *Bee Craft* 91, no. 2: 4–6.

Chapman, H. A., and A. K. Anderson. 2012. Understanding disgust. *Annals of the New York Academy of Sciences* 1251: 62–76.

Chechetka, S. A., Y. Yu, M. Tange, and E. Miyako. 2017. Materially engineered artificial pollinators. *Chem* 2: 234–239.

Chittka, L., A. Schmida, N. Troje, and R. Menzel. 1994. Ultraviolet as a component of flower reflections, and the color perception of Hymenoptera. *Vision Research* 34: 1489–1508.

Chittka, L., and N. M. Wasser. 1997. Why red flowers are not invisible to bees. *Israel Journal of Plant Sciences* 45: 169–183.

Clarke, D., H. Whitney, G. Sutton, and D. Robert. 2013. Detection and learning of floral electric fields by bumblebees. *Science* 340: 66–69.

Cnaani, J., J. D. Thomson, and D. R. Papaj. 2006. Flower choice and learning in foraging bumblebees: Effects of variation in nectar volume and concentration. *Ethology* 112: 278–285.

Code, B. H., and S. L. Haney. 2006. Franklin's bumble bee inventory in the southern Cascades of Oregon. Medford, OR: Bureau of Land Management, 8 pp.

Coleridge, S. 1853. *Pretty Lessons in Verse for Good Children, with Some Lessons in Latin in Easy Rhyme*. London: John W. Parker and Son.

Correll, D. S. 1953. Vanilla: Its botany, history, cultivation and economic import. *Economic Botany* 7: 291–358.

Crane, E. 1999. *The World History of Beekeeping and Honey Hunting*. New York: Routledge.

Crepet, W. L., and K. C. Nixon. 1998. Fossil Clusiaceae from the late Cretaceous (Turonian) of New Jersey and implications regarding the history of bee pollination. *American Journal of Botany* 85: 1122–1133.

Crittenden, A. N. 2011. The importance of honey consumption in human evolution. *Food and Foodways* 19: 257–273.

———. 2016. Ethnobotany in evolutionary perspective: Wild plants in diet composition and daily use among Hadza hunter-gatherers. Pp. 319–340 in K. Hardy and L. Kubiak-Martens, eds., *Wild Harvest: Plants in the Hominin and Pre-Agrarian Human Worlds*. Oxford: Oxbow Books.

Crittenden, A. N., N. L. Conklin-Britain, D. A. Zes, M. J. Schoeninger, et al. 2013. Juvenile foraging among the Hadza: Implications for human life history. *Evolution and Human Behavior* 34: 299–304.

Crittenden, A. N., and S. L. Schnorr. 2017. Current views on hunter-gatherer nutrition and the evolution of the human diet. *Yearbook of Physical Anthropology* 162(S63): 84–109.

Crittenden, A. N., and D. A. Zess. 2015. Food sharing among Hadza hunter-gatherer children. *PLoS ONE* 10: e0131996.

Cutler, G. C., C. D. Scott-Dupree, M. Sultan, A. D. McFarlane, et al. 2014. A large-scale field study examining effects of exposure to clothianidin seed-treated canola on honey bee colony health, development, and overwintering success. *PeerJ* 2: e652. https://doi.org/10.7717/peerj.652.

D'Andrea, L., F. Felber, and R. Guadagnulo. 2008. Hybridization rates between lettuce (*Lactuca sativa*) and its wild relative (*L. serriola*) under field conditions. *Environmental Biosafety Research* 7: 61–71.

Danforth, B. N. 1999. Emergence, dynamics, and bet hedging in a desert mining bee, *Perdita portalis*. *Proceedings of the Royal Society B* 266: 1985–1994.

———. 2002. Evolution of sociality in a primitively eusocial lineage of bees. *Proceedings of the National Academy of Sciences* 99: 286–290.

Danforth, B. N., S. Cardinal, C. Praz, E. A. B. Almeida, et al. 2013. The impact of molecular data on our understanding of bee phylogeny and evolution. *Annual Review of Entomology* 58: 57–78.

Danforth, B. N., and G. O. Poinar, Jr. 2011. Morphology, classification, and antiquity of *Melittosphex burmensis* (Apoidea: Melittosphecidae) and implications for early bee evolution. *Journal of Paleontology* 85: 882–891.

Danforth, B. N., S. Sipes, J. Fang, and S. G. Brady. 2006. The history of early bee diversification based on five genes plus morphology. *Proceedings of the National Academy of Sciences* 103: 15118–15123.

Darwin, C. 1859. *On the Origin of Species by Means of Natural Selection.* (Reprint of 1859 first edition.) Mineola, NY: Dover.

———. 1877. *The Various Contrivances by Which Orchids Are Fertilised by Insects*, 2nd ed. New York: D. Appleton and Company.

Dean, W. R. J., W. R. Siegfried, and I. A. W. MacDonald. 1990. The fallacy, fact, and fate of guiding behavior in the Greater Honeyguide. *Conservation Biology* 4: 99–101.

Dicks, L. V., B. Viana, R. Bommarco, B. Brosi, et al. 2016. Ten policies for pollinators. *Science* 354: 975–976.

Dillon, M. E., and R. Dudley. 2014. Surpassing Mt. Everest: Extreme flight performance of alpine bumble-bees. *Biology Letters* 10. https://doi.org/10.1098/rsbl.2013.0922.

Di Prisco, G., D. Annoscia, M. Margiotta, R. Ferrara, et al. 2016. A mutualistic symbiosis between a parasitic mite and a pathogenic virus undermines honey bee immunity and health. *Proceedings of the National Academy of Sciences* 113: 3203–3208.

Doyle, A. C. 1917. *His Last Bow: A Reminiscence of Sherlock Holmes.* New York: Review of Reviews Company.

Doyle, J. A. 2012. Molecular and fossil evidence on the origin of angiosperms. *Annual Review of Earth and Planetary Sciences* 40: 301–326.

Driscoll, C. A., D. W. Macdonald, and S. J. O'Brian. 2009. From wild animals to domestic pets, an evolutionary view of domestication. *Proceedings of the National Academy of Sciences* 106: 9971–9978.

Eckert, J. E. 1933. The flight range of the honeybee. *Journal of Agricultural Research* 47: 257–286.

Eilers E. J., C. Kremen, S. Smith Greenleaf, A. K. Garber, et al. 2011. Contribution of pollinator-mediated crops to nutrients in the human food supply. *PLoS ONE* 6: e21363. https://doi.org/10.1371/journal.pone.0021363.

Engel, M. S. 2000. A new interpretation of the oldest fossil bee (Hymenoptera: Apidae). *American Museum Novitiates*, no. 3296, 11 pp.

———. 2001. *A monograph of the Baltic amber bees and evolution of the apoidea (Hymenoptera). Bulletin of the American Museum of Natural History* 259, 192 pp.

Escobar, T. 2007. *Curse of the Nemur: In Search of the Art, Myth, and Ritual of the Ishir.* Pittsburgh: University of Pittsburgh Press.

Evangelista, C., P. Kraft, M. Dacke, J. Reinhard, et al. 2010. The moment before touchdown: Landing manoeuvres of the honeybee *Apis mellifera. Journal of Experimental Biology* 213: 262–270.

Evans, E., R. Thorp, S. Jepsen, and S. H. Black. 2008. *Status Review of Three Formerly Common Species of Bumble Bee in the Subgenus* Bombus. Portland, OR: Xerces Society for Invertebrate Conservation, 63 pp.

Evans, H. E., and K. M. O'Neill. 2007. *The Sand Wasps: Natural History and Behavior.* Cambridge, MA: Harvard University Press.

Fabre, J. E. 1915. *Bramble-Bees and Others.* New York: Dodd, Mead.

———. 1916. *The Mason-Bees.* New York: Dodd, Mead.

Fenster, C. B., W. X. Armbruster, P. Wilson, M. R. Dudash, et al. 2004. Pollination syndromes and floral specialization. *Annual Review of Ecology, Evolution, and Systematics* 35: 375–403.

Filella, I., J. Bosch, J. Llusià, R. Seco, et al. 2011. The role of frass and cocoon volatiles in host location by *Monodontomerus aeneus*, a parasitoid of Megachilid solitary bees. *Environmental Entomology* 40: 126–131.

Fine, J. D., D. L. Cox-Foster, and C. A. Mullein. 2017. An inert pesticide adjuvant synergizes viral pathogenicity and mortality in honey bee larvae. *Scientific Reports* 7. https://doi.org/10.1038/srep40499.

Friedman, W. E. 2009. The meaning of Darwin's "Abominable Mystery." *American Journal of Botany* 96: 5–21.

Friis, E. M., P. R. Crane, and K. R. Pedersen. 2011. *Early Flowers and Angiosperm Evolution.* Cambridge: Cambridge University Press.

Garibaldi, L. A., I. Steffan-Dewenter, R. Winfree, M. A. Aizen, et al. 2013. Wild pollinators enhance fruit set of crops regardless of honey bee abundance. *Science* 339: 1608–1611.

Gegear, R. J., and J. G. Burns. 2007. The birds, the bees, and the virtual flowers: Can pollinator behavior drive ecological speciation in flowering plants? *American Naturalist* 170. https://doi.org/10.1086/521230.

Genersch, E., C. Yue, I. Fries, and J. R. de Miranda. 2006. Detection of *Deformed wing virus*, a honey bee viral pathogen, in bumble bees (*Bombus terrestris* and *Bombus pascuorum*) with wing deformities. *Journal of Invertebrate Pathology* 91: 61–63.

Gess, S. K. 1996. *The Pollen Wasps: Ecology and Natural History of the Masarinae.* Cambridge, MA: Harvard University Press.

Gess, S. K., and F. W. Gess. 2010. *Pollen Wasps and Flowers in Southern Africa.* Pretoria: South African National Biodiversity Institute.

Ghazoul, J. 2005. Buzziness as usual? Questioning the global pollination crisis. *TRENDS in Ecology and Evolution* 20: 367–373.

Glaum, P., M. C. Simayo, C. Vaidya, G. Fitch, et al. 2017. Big city Bombus: Using natural history and land-use history to find significant environmental drivers in bumble-bee declines in urban development. *Royal Society Open Science* 4: 170156.

Goor, A. 1967. The history of the date through the ages in the Holy Land. *Economic Botany* 21: 320–340.

Goubara, M., and T. Takasaki. 2003. Flower visitors of lettuce under field and enclosure conditions. *Applied Entomology and Zoology* 38: 571–581.

Goulson, D. 2010. Impacts of non-native bumblebees in Western Europe and North America. *Applied Entomology and Zoology* 45: 7–12.

Goulson, D., E. Nicholls, C. Botías, and E. L. Rotheray. 2015. Bee declines driven by combined stress from parasites, pesticides, and lack of flowers. *Science* 347. https://doi.org/10.1126/science.1255957.

Goulson, D., and J. C. Stout. 2001. Homing ability of the bumblebee *Bombus terrestris* (Hymenoptera: Apidae). *Apidologie* 32: 105–111.

Graves, R. 1960. *The Greek Myths.* London: Penguin.

Greceo, M. K., P. M. Welz, M Siegrist, S. J. Ferguson, et al. 2011. Description of an ancient social bee trapped in amber using diagnostic radioentomology. *Insectes Sociaux* 58: 487–494.

Griffin, B. 1997a. *Humblebee Bumblebee.* Bellingham, WA: Knox Cellars Publishing.

———. 1997b. *The Orchard Mason Bee.* Bellingham, WA: Knox Cellars Publishing.

Grimaldi, D. 1996. *Amber: Window to the Past.* New York: Harry N. Abrams.

———. 1999. The co-radiations of pollinating insects and angiosperms in the Cretaceous. *Annals of the Missouri Botanical Garden* 86: 373–406.

Grimaldi, D., and M. Engel. 2005. *Evolution of the Insects.* New York: Cambridge University Press.

Hallmann, C. A., R. P. B. Foppen, C. A. M. van Turnhout, H. de Kroon, et al. 2014. Declines in insectivorous birds are associated with high neonicotinoid concentrations. *Nature* 511: 341–343.

Hanson, T., and J. S. Ascher. 2018. An unusually large nesting aggregation of the digger bee *Anthophora bomboides* Kirby, 1838 (Hymenoptera: Apidae) in the San Juan Islands, Washington State. *Pan-Pacific Entomologist* 94: 4-16.

Hedtke, S. M., S. Patiny, and B. N. Danorth. 2013. The bee tree of life: A supermatrix approach to apoid phylogeny and biogeography. *BMC Evolutionary Biology* 13: 138.

Heinrich, B. 1979. *Bumblebee Economics.* Cambridge, MA: Harvard University Press.

Henderson, A. 1986. A review of pollination studies in the Palmae. *Botanical Review* 52: 221–259.

Herodotus. 1997. *The Histories.* Translated by G. Rawlinson. New York: Knopf.

Hershorn, C. 1980. Cosmetics queen Mary Kay delivers a megabuck message to her sales staff: 'Women can do anything.' *People,* http://people .com/archive/cosmetics-queen-mary-kay-delivers-a-megabuck -message-to-her-sales-staff-women-can-do-anything-vol-13-no-17.

Hoballah, M. E., T. Gübitz, J. Stuurman, L. Broger, et al. 2007. Single gene-mediated shift in pollinator attraction in Petunia. *Plant Cell* 19: 779–790.

Hogue, C. L. 1987. Cultural entomology. *Annual Review of Entomology* 32: 181–199.

Houston, T. F. 1984. Biological observations of bees in the genus *Ctenocolletes* (Hymenoptera: Stenotritidae). *Records of the Western Australian Museum* 11: 153–172.

How, M. J., and J. M. Zanker. 2014. Motion camouflage induced by zebra stripes. *Zoology* 117: 163–170.

Ichikawa, M. 1981. Ecological and sociological importance of honey to the Mbuti net hunters, Eastern Zaire. *African Study Monographs* 1: 55–68.

Iwasa, T., N. Motoyama, J. T. Ambrose, and R. M. Roe. 2004. Mechanism for the differential toxicity of neonicotinoid insecticides in the honey bee, *Apis mellifera*. *Crop Protection* 23: 371–378.

Jablonski, P. G., H. J. Cho, S. R. Song, C. K. Kang, et al. 2013. Warning signals confer advantage to prey in competition with predators: Bumblebees steal nests from insectivorous birds. *Behavioral Ecology and Sociobiology* 67: 1259–1267.

Jacob, F. 1977. Evolution and tinkering. *Science* 196: 1161–1166.

Jones, H. A. 1927. Pollination and life history studies of lettuce (*Lactuca sativa* L.). *Hilgardia* 2: 425–479.

Jones, K. N., and J. S. Reithel. 2001. Pollinator-mediated selection on a flower color polymorphism in experimental populations of *Antirrhinum* (Scrophulariaceae). *American Journal of Botany* 88: 447–454.

Kajobe, R., and D. W. Roubik. 2006. Honey-making bee colony abundance and predation by apes and humans in a Uganda forest reserve. *Biotropica* 38: 210–218.

Kerr, J. T., A. Pindar, P. Galpern, L Packer, et al. 2015. Climate change impacts on bumblebees converge across continents. *Science* 349: 177–180.

Kevan, P. G., L. Chittka, and A. G. Dyer. 2001. Limits to the salience of ultraviolet: Lessons from colour vision in bees and birds. *Journal of Experimental Biology* 204: 2571–2580.

Keynes, R., ed. 2010. *Charles Darwin's Zoology Notes and Specimen Lists from H.M.S. Beagle*. Cambridge: Cambridge University Press.

Kirchner, W. H., and J. Röschard. 1999. Hissing in bumblebees: An interspecific defence signal. *Insectes Sociaux* 46: 239–243.

Klein, A., C. Brittain, S. D. Hendrix, R. Thorp, et al. 2012. Wild pollination services to California almond rely on semi-natural habitat. *Journal of Applied Ecology* 49: 723–732.

Klein, A., B. E. Vaissière, J. H. Cane, I. Steffan-Dewenter, et al. 2007. Importance of pollinators in changing landscapes for world crops. *Proceedings of the Royal Society B* 274: 303–313.

Koch, J. B., and J. P. Strange. 2012. The status of *Bombus occidentalis* and *B. moderatus* in Alaska with special focus on *Nosema bombi* incidence. *Northwest Science* 86: 212–220.

Kritsky, G. 1991. Darwin's Madagascan hawk moth prediction. *American Entomologist* 37: 205–210.

Krombein, K., and B. Norden. 1997a. Bizarre nesting behavior of *Krombeinictus nordenae* Leclercq (Hymenoptera: Sphecidae, Crabroninae). *Journal of South Asian Natural History* 2: 145–154.

———. 1997b. Nesting behavior of *Krombeinictus nordenae* Leclercq, a sphecid wasp with vegetarian larvae (Hymenoptera: Sphecidae, Crabroninae). *Proceedings of the Entomological Society of Washington* 99: 42–49.

Krombein, K. V., B. B. Norden, M. M. Rickson, and F. R. Rickson. 1999. Biodiversity of the Domatia occupants (ants, wasps, bees and others) of the Sri Lankan Myrmecophyte *Humboldtia lauifolia* (Fabaceae). *Smithsonian Contributions to Zoology* 603: 1–34.

Larison B., R. J. Harrigan, H. A. Thomassen, D. I. Rubenstein, et al. 2015. How the zebra got its stripes: A problem with too many solutions. *Royal Society Open Science* 2: 140452.

Larue-Kontić, A. C., and R. R. Junker. 2016. Inhibition of biochemical terpene pathways in *Achillea millefolim* flowers differently affects the behavior of bumblebees (*Bombus terrestris*) and flies (*Lucilia sericata*). *Journal of Pollination Ecology* 18: 31–35.

Lee, D. 2007. *Nature's Palette: The Science of Plant Color*. Chicago: University of Chicago Press.

Lewis-Williams, J. D. 2002. *A Cosmos in Stone: Interpreting Religion and Society Through Rock Art*. Walnut Creek, CA: AltaMira Press.

Linnaeus, C. 1737. *Critica Botanica*. Leiden: Conradum Wishoff.

Litman, J. R., B. N. Danforth, C. D. Eardley, and C. J. Praz. 2011. Why do leafcutter bees cut leaves? New insights into the early evolution of bees. *Proceedings of the Royal Society B* 278: 3593–3600.

Livy. 1938. *The History of Rome*, Books 40–42. Translated by E. T. Sage and A. C. Schlesinger. Cambridge, MA: Harvard University Press. Archived online at Perseus Digital Library, Tufts University, www.perseus.tufts.edu/hopper.

Lockwood, J. 2013. *The Infested Mind: Why Humans Fear, Loathe, and Love Insects*. New York: Oxford University Press.

Longfellow, H. W. 1893. *The Complete Poetical Works of Henry Wadsworth Longfellow*. Boston: Houghton Mifflin.

Lucano, M. J., G. Cernicchiaro, E. Wajnberg, and D. M. S. Esquivel. 2005. Stingless bee antennae: A magnetic sensory organ? *BioMetals* 19: 295–300.

Lunau, K. 2004. Adaptive radiation and coevolution—Pollination biology case studies. *Organisms, Diversity & Evolution* 4: 207–224.

Maeterlinck, M. 1901. *The Life of Bees*. Translated by A. Sutro. Cornwall, NY: Cornwall Press.

Marlowe, F. W., J. C. Berbesque, B. Wood, A. Crittenden, et al. 2014. Honey, Hadza, hunter-gatherers, and human evolution. *Journal of Human Evolution* 71: 119–128.

McGovern, P., J. Zhang, J. Tang, Z. Zhang, et al. 2004. Fermented beverages of pre- and proto-historic China. *Proceedings of the National Academy of Sciences* 101: 17593–17598.

McGregor, S. E. 1976. *Insect Pollination of Cultivated Crop Plants*. USDA Agriculture Handbook no. 496. Updated version available at US Department of Agriculture, Agricultural Research Service, http://gears.tucson.ars.ag.gov/book.

Messer, A. C. 1984. *Chalicodoma pluto*: The world's largest bee rediscovered living communally in termite nests (Hymenoptera: Megachilidae). *Journal of the Kansas Entomological Society* 57: 165–168.

Meyer, R. S., A. E. DuVal, and H. R. Jensen. 2012. Patterns and processes in crop domestication: An historical review and quantitative analysis of 203 global food crops. *New Phytologist* 196: 29–48.

Michener, C. D. 2007. *The Bees of the World*. Baltimore: Johns Hopkins University Press.

Michener, C. D., and D. A. Grimaldi. 1988. The oldest fossil bee: Apoid history, evolutionary stasis, and antiquity of social behavior. *Proceedings of the National Academy of Sciences* 85: 6424–6426.

Miller, W. 1955. Old man's advice to youth: Never lose your curiosity. *Life*, May 2, 62–64.

Mobbs, D., R. Yu, J. B. Rowe, H. Eich, et al. 2010. Neural activity associated with monitoring the oscillating threat value of a tarantula. *Proceedings of the National Academy of Sciences* 107: 20582–20586.

Moritz, R. F. A., and R. M. Crewe. 1988. Air ventilation in nests of two African stingless bees *Trigona denoiti* and *Trigona gribodoi*. *Experientia* 44: 1024–1027.

Muir, J. 1882a. The bee-pastures of California, Part I. *Century Magazine* 24: 222–229.

———. 1882b. The bee-pastures of California, Part II. *Century Magazine* 24: 388–395.

Mullin, C. A., M. Frazier, J. L. Frazier, S. Ashcraft, et al. 2010. High levels of miticides and agrochemicals in North American apiaries: Implications for honey bee health. *PLoS ONE* 5: e9754. https://doi.org/10.1371/journal.pone.0009754.

Nichols, W. J. 2014. *Blue Mind*. New York: Little, Brown.

Nininger, H. H. 1920. Notes on the life-history of *Anthophora stanfordiana*. *Psyche* 27: 135–137.

O'Neill, K. M. 2001. *Solitary Wasps: Behavior and Natural History*. Ithaca, NY: Cornell University Press.

Ott, J. 1998. The Delphic bee: Bees and toxic honeys as pointers to psychoactive and other medicinal plants. *Economic Botany* 52: 260–266.

Packer, L. 2005. A new species of *Geodiscelis* (Hymenoptera: Colletidae: Xeromelissinae) from the Atacama Desert of Chile. *Journal of Hymenoptera Research* 14: 84–91.

Paris, H. S., and J. Janick. 2008. What the Roman emperor Tiberius grew in his greenhouses. Pp. 33–41 in M. Pitrat, ed., *Cucurbitaceae 2008: Proceedings of the IXth EUCARPIA Meeting on Genetics and Breeding of Cucurbitaceae*. Avignon, France: INRA.

Partap, U., and T. Ya. 2012. The human pollinators of fruit crops in Maoxian County, Sichuan, China. *Mountain Research and Development* 32: 176–186.

Peckham, G. W., and E. G. Peckham. 1905. *Wasps: Social and Solitary*. Boston: Houghton Mifflin.

Phillips, E. F. 1905. Structure and development of the compound eye of the honeybee. *Proceedings of the Academy of Natural Sciences of Philadelphia* 56: 123–157.

Plath, O. E. 1934. *Bumblebees and Their Ways*. New York: Macmillan.

Plath, S. 1979. *Johnny Panic and the Bible of Dreams*. New York: Harper and Row.

Poinar, G. O., Jr., K. L. Chambers, and J. Wunderlich. 2013. *Micropetasos*, a new genus of angiosperms from mid-Cretaceous Burmese amber. *Journal of the Botanical Research Institute of Texas* 7: 745–750.

Poinar, G. O., Jr., and B. N. Danforth. 2006. A fossil bee from early Cretaceous Burmese amber. *Science* 314: 614.

Poinar, G. O., Jr., and R. Poinar. 2008. *What Bugged the Dinosaurs: Insects, Disease and Death in the Cretaceous*. Princeton, NJ: Princeton University Press.

Porter, C. J. A. 1883. Experiments with the antennae of insects. *American Naturalist* 17: 1238–1245.

Porter, D. M. 2010. Darwin: The botanist on the *Beagle*. *Proceedings of the California Academy of Sciences* 61: 117–156.

Potts, S. G., J. C. Biesmeijer, C. Kremen, P. Neumann, et al. 2010. Global pollinator declines: Trends, impacts and drivers. *Trends in Ecology & Evolution* 25: 345–353.

Potts, S. G., V. L. Imperatriz-Fonseca, and H. T. Ngo, eds. 2016. *The Assessment Report of the Intergovernmental Science-Policy Platform on Biodiversity and Ecosystem Services on Pollinators, Pollination and Food Production*. Bonn, Germany: Secretariat of the Intergovernmental Science-Policy Platform on Biodiversity and Ecosystem Services.

Proctor, M., P. Yeo, and A. Lack. 1996. *The Natural History of Pollination*. Portland, OR: Timber Press.

Pyke, G. H. 2016. Floral nectar: Pollination attraction or manipulation? *Trends in Ecology and Evolution* 31: 339–341.

Ransome, H. M. 2004. *The Sacred Bee in Ancient Times and Folklore*. (Reprint of 1937 edition.) Mineola, NY: Dover.

Reinhardt, J. F. 1952. Some responses of honey bees to alfalfa flowers. *American Naturalist* 86: 257–275.

Roffet-Salque, M., M. Regert, R. P. Evershed, A. K. Outram, et al. 2015. Widespread exploitation of the honeybee by early Neolithic farmers. *Nature* 527: 226–231.

Ross, A., C. Mellish, P. York, and B. Crighton. 2010. Burmese amber. Pp. 208–235 in D. Penny, ed., *Biodiversity of Fossils in Amber from the Major World Deposits*. Manchester, UK: Siri Scientific Press.

Roubik, D. W., ed. 1995. *Pollination of Cultivated Plants in the Tropics*. Rome: Food and Agriculture Organization of the United Nations.

Roulston, T., and K. Goodell. 2011. The role of resources and risks in regulating wild bee populations. *Annual Review of Entomology* 56: 293–312.

Rundlöf, M., G. K. S. Andersson, R. Bommarco, I. Fries, et al. 2015. Seed coating with a neonicotinoid insecticide negatively affects wild bees. *Nature* 521: 77–80.

Saunders, E. 1896. *The Hymenoptera Aculeata of the British Islands*. London: L. Reeve.

Savage, C. 2008. *Bees: Natures Little Wonders*. Vancouver, BC: Greystone Books.

Schemske, D. W., and H. D. Bradshaw, Jr. 1999. Pollinator preference and the evolution of floral traits in monkeyflowers (Mimulus). *Proceedings of the National Academy of Sciences* 96: 11910–11915.

Schmidt. J. O. 2014. Evolutionary responses of solitary and social Hymenoptera to predation by primates and overwhelmingly powerful vertebrate predators. *Journal of Human Evolution* 71: 12–19.

———. 2016. *The Sting of the Wild.* Baltimore: Johns Hopkins University Press.

Schwarz, H. F. 1945. The wax of stingless bees (Meliponidæ) and the uses to which it has been put. *Journal of the New York Entomological Society* 53: 137–144.

Schwarz, M. P., M. H. Richards, and B. N. Danforth. 2007. Changing paradigms in insect social evolution: Insights from halictine and allodapine bees. *Annual Review of Entomology* 52: 127–150.

Seligman, M. E. P. 1971. Phobias and preparedness. *Behavior Therapy* 2: 307–320.

Shackleton, K., H. A. Toufailia, N. J. Balfour, F. S. Nasicimento, et al. 2015. Appetite for self-destruction: Suicidal biting as a nest defense strategy in *Trigona* stingless bees. *Behavioral Ecology and Sociobiology* 69: 273–281.

Slaa, E. J., L. Alejandro, S. Chaves, K. Sampaio Malagodi-Braga, et al. 2006. Stingless bees in applied pollination: Practice and perspectives. *Apidologie* 37: 293–315.

Sladen, F. W. L. 1912. *The Humble-Bee: Its Life-History and How to Domesticate It.* London: Macmillan.

Smith, A. 2012. Cash-strapped farmers feed candy to cows. CNN Money, http://money.cnn.com/2012/10/10/news/economy/farmers-cows-candy-feed/index.html.

Somanathan, H., A. Kelber, R. M. Borges, R. Wallén, et al. 2009. Visual ecology of Indian carpenter bees II: Adaptations of eyes and ocelli to nocturnal and diurnal lifestyles. *Journal of Comparative Physiology A* 195: 571–583.

Sparrman, A. 1777. An account of a journey into Africa from the Cape of Good-Hope, and a description of a new species of cuckow. In a letter to Dr. John Reinhold Forster, FRS *Philosophical Transactions of the Royal Society of London* 67: 38–47.

Srinivasan, M. V. 1992. Distance perception in insects. *Current Directions in Psychological Science* 1: 22–26.

Stableton, J. K. 1908. Observation beehive. *School and Home Education* 28: 21–23.

Stokstad, E. 2007. The case of the empty hives. *Science* 316: 970–972.

Stone, G. N. 1993. Endothermy in the solitary bee *Anthophora plumipes*: Independent measures of thermoregulatory ability, costs of warm-up and the role of body size. *Journal of Experimental Biology* 174: 299–320.

Strong, D. R., J. H. Lawton, and R. Southwood. 1984. *Insects on Plants: Community Patterns and Mechanisms.* Cambridge, MA: Harvard University Press.

Sun, B. Y., T. F. Stuess, A. M. Humana, M. Riveros, et al. 1996. Evolution of *Rhaphithamnus venustus* (Verbenaceae), a gynodioecious hummingbird-pollinated endemic of the Juan Fernandez Islands, Chile. *Pacific Science* 50: 55–65.

Sutherland, W. J. 1990. Biological flora of the British Isles: *Iris pseudacorus* L. *Journal of Ecology* 78: 833–848.

Theophrastus. 1916. *Enquiry into Plants, and Minor Works on Odours and Weather Signs.* Translated by A. Hort. London: William Heinemann.

Thoreau, H. D. 1843. Paradise (to be) regained. *United States Magazine and Democratic Review* 13: 451–463.

———. 2009. *The Journal, 1837–1861.* Edited by D. Searls. New York: New York Review Books.

Thorp, R. W. 1969. Ecology and behavior of *Anthophora edwardsii. American Midland Naturalist* 82: 321–337.

Tolstoy, L. (1867) 1994. *War and Peace.* New York: Modern Library.

Torchio, P. F. 1984. The nesting biology of *Hylaeus bisinuatus* Forster and development of its immature forms (Hymenoptera: Colletidae). *Journal of the Kansas Entomological Society* 57: 276–297.

Torchio, P. F., and V. J. Tepedino. 1982. Parsivoltinism in three species of *Osmia* bees. *Psyche* 89: 221–238.

VanEngelsdorp, D., D. Cox-Foster, M. Frazier, N. Ostiguy, et al. 2006. "Fall-Dwindle Disease": Investigations into the causes of sudden and alarming colony losses experienced by beekeepers in the fall of 2006. Mid-Atlantic Apiculture Research and Extension Consortium (MAAREC)–Colony Collapse Disorder Working Group, 22 pp.

VanEngelsdorp, D., J. D. Evans, L. Donovall, C. Mullin, et al. 2009. "Entombed Pollen": A new condition in honey bee colonies associated with increased risk of colony mortality. *Journal of Invertebrate Pathology* 101: 147–149.

Virgil. 2006. *The Georgics.* Translated by P. Fallon. Oxford: Oxford University Press.

Wallace, Alfred Russel. 1869. *The Malay Archipelago.* New York: Harper and Brothers.

Watson, K., and J. A. Stallins. 2016. Honey bees and Colony Collapse Disorder: A pluralistic reframing. *Geography Compass* 10: 222–236.

Wcislo, W. T., L Arneson, K. Roesch, V. Gonzolez, et al. 2004. The evolution of nocturnal behaviour in sweat bees, *Megalopta genalis* and M. *ecuadoria* (Hymenoptera: Halictidae): An escape from competitors and enemies? *Biological Journal of the Linnean Society* 83: 377–387.

Wcislo, W. T., and B. N. Danforth. 1997. Secondarily solitary: The evolutionary loss of social behavior. *Trends in Ecology and Evolution* 12: 468 474.

Wellington, W. G. 1974. Bumblebee ocelli and navigation at dusk. *Science* 183: 550–551.

Weyrich, L. S., S. Duchene, J. Soubrier, L. Arriola, et al. 2017. Neanderthal behaviour, diet, and disease inferred from ancient DNA in dental calculus. *Nature* 544: 357–361.

Whitfield, C. W., S. K. Behura, S. H. Berlocher, A. G. Clark, et al. 2007. Thrice out of Africa: Ancient and recent expansions of the honey bee, *Apis mellifera*. *Science* 314: 642–645.

Whitman, W. (1855) 1976. *Leaves of Grass*. Secaucus, NJ: Longriver Press.

Whitney, H. M., L. Chittka, T. J. A. Bruce, and B. J. Glover. 2009. Conical epidermal cells allow bees to grip flowers and increase foraging efficiency. *Current Biology* 19: 948–953.

Wille, A. 1983. Biology of the stingless bees. *Annual Review of Entomology* 28: 41–64.

Wilson, E. O. 2012. *The Social Conquest of Earth*. New York: Liveright.

Winston, M. L. 1987. *The Biology of the Honey Bee*. Cambridge, MA: Harvard University Press.

Wood, B. M., H. Pontzer, D. A. Raichlen, and F. W. Marlowe. 2014. Mutualism and manipulation in Hadza-honeyguide interactions. *Evolution and Human Behavior* 35: 540–546.

Wrangham, R. W. 2011. Honey and fire in human evolution. Pp. 149–167 in J. Sept and D. Pilbeam, eds. *Casting the Net Wide: Papers in Honor of Glynn Isaac and His Approach to Human Origins Research*. Oxford: Oxbow Books.

Yeats. W. B. 1997. *The Collected Works of W. B. Yeats*. Vol. 1, *The Poems*, 2nd ed. Edited by J. Finneman. New York: Scribner.

Index

Thor Hanson is a conservation biologist, Guggenheim fellow, and author of award-winning books including *Feathers*, *The Impenetrable Forest*, and *The Triumph of Seeds*. He lives with his wife and son on an island in Washington State.

www.thorhanson.net